When Machines Decide: Building Moral Architectures for an AI-Human World

By: Alexandra Isaacs

2026

COPYRIGHT

This book is a work of nonfiction. All research, examples, and references are presented for educational purposes. While every effort has been made to ensure accuracy, the author and publisher assume no responsibility for errors or omissions. For permissions, inquiries, or collaborations, please contact:

ISBN: 978-1-970503-31-9
Published by: WDP Publishing
Third Edition of The Future Series 2026

Visit the author's page on Amazon for more information and updates:
Alexandra Isaacs

ABOUT THE AUTHOR

ALEXANDRA ISAACS

Alexandra Isaacs is a researcher whose work confronts one of the defining questions of our era: Will artificial intelligence elevate humanity or endanger it? Rooted in health technology and human-centered design, she has seen firsthand how powerful tools can heal, empower, and transform lives… and how easily they can be misused.

This duality, innovation's capacity for both good and harm, fuels her mission as an author. Alexandra writes not only to inform but also to influence: to shift public understanding, inspire readers to ask deeper questions, and encourage leaders to adopt principles that safeguard human dignity and freedom.

Alexandra writes to help readers understand the moral crossroads we now stand upon. Her work examines AI not just as a technological advancement but as a force that reflects human values, ethics, and intentions. Through clear analysis and balanced insight, she reveals how AI can be harnessed for compassion, fairness, and global wellbeing, or twisted toward manipulation, exploitation, and control.

Her books invite readers to become informed participants in shaping the future, reminding us that the true battle between good and evil will not be fought by machines, but by the humans who build and guide them

ABSTRACT

We stand at a crossroads where the decisions we make about artificial intelligence will fundamentally shape the future of human agency, accountability, and dignity. This book is not a warning about machines taking over, nor is it a celebration of technological salvation. Instead, it is an invitation to engage deeply with one of the most consequential questions of our time: How do we design AI systems that enhance rather than diminish our humanity?

Across boardrooms, government offices, research labs, and everyday workplaces, AI is already making decisions that affect lives, livelihoods, and the fabric of society. Algorithms determine who gets hired, who receives medical treatment, which investments succeed, and how resources are allocated. Yet most of us—even those deploying these systems—struggle to answer fundamental questions about accountability, ethics, and control.

This book emerges from years of working at the intersection of technology and human-centered design and witnessing both the extraordinary potential of AI to heal and empower, and its capacity to harm when deployed without adequate ethical frameworks. What i've learned is that the battle between good and evil in the age of AI will not be fought by machines. It will be fought by the humans who build, deploy, and govern them.

You don't need to be a data scientist or philosopher to participate in shaping this future. Whether you're a corporate strategist navigating digital transformation, a policymaker crafting governance framework, an entrepreneur seeking competitive advantage, or a professional concerned about automation's impact on your career, this book is for you. It provides practical frameworks for understanding AI decision-making, evaluating its ethical implications, and implementing responsible systems that preserve human values while capturing technological benefits.

Each chapter progresses from familiar territory to more complex considerations; while grounding abstract concepts in real-world scenarios you might encounter in your work. We'll explore how human responsibility transforms (rather than disappears) when AI makes decisions, examine frameworks for determining when to trust algorithmic judgment, and develop practical approaches for maintaining accountability in AI-augmented organizations.

Most importantly, this book emphasizes collaboration. Responsible AI governance cannot be achieved by technologists alone, nor by ethicists, business leaders, or policymakers working in isolation. It requires all of us across various disciplines and perspectives to engage in thoughtful dialogue about what we value and how we want to live.

The future is not predetermined. The choices we make today about AI design, deployment, and governance will echo for generations. This book equips you to make those choices wisely, ensuring that the machines we build serve humanity's highest aspirations rather than its basest impulses. Let's begin.

TABLE OF CONTENTS

INTRODUCTION

THe delegation of decision-making authority to artificial intelligence systems represents one of the most consequential organizational and societal shifts of the twenty-first century. Across industries and institutions, algorithms now influence choices that affect millions of lives, determining creditworthiness, recommending medical treatments, allocating resources, optimizing supply chains, and shaping strategic business directions. This transformation extends far beyond simple automation of routine tasks. Ai systems increasingly operate as collaborative partners in decision-making processes by identifying patterns humans cannot perceive, processing information at scales that exceed human capacity, and generating recommendations that executives, policymakers, and professionals must evaluate and act upon.

The central challenge this transition creates is not primarily technical. Organizations have successfully deployed sophisticated algorithms, and the technology continues advancing rapidly. The fundamental question concerns accountability, governance, and the preservation of human agency in environments where machines possess superior analytical capabilities in specific domains. When an AI system recommends a course of action that contradicts human judgment, who decides which perspective prevails? When algorithmic decisions produce harmful outcomes, who bears responsibility? When competitive pressures incentivize automation, how do organizations maintain meaningful human oversight without sacrificing efficiency advantages?

These questions demand urgent attention because AI decision-making systems are already embedded in critical infrastructure, not hypothetical future scenarios. Healthcare providers rely on algorithms to prioritize patient care. Financial institutions use AI to assess risk and allocate capital. Governments deploy automated systems to distribute benefits and identify fraud. Corporations integrate machine learning into strategic planning and operational management. Each implementation creates new accountability challenges that existing governance frameworks struggle to address.

There is added complexity because AI decision-making defies simple categorization as either beneficial or dangerous. The same algorithmic capabilities that enable breakthrough medical diagnoses can perpetuate historical biases in treatment recommendations. The pattern recognition that optimizes supply chains and reduces costs can also identify efficiencies that eliminate jobs and destabilize communities. The predictive analytics that help businesses anticipate customer needs can manipulate consumer behavior in ethically questionable ways. Responsible integration of AI requires navigating these tensions rather than embracing technological determinism or rejecting innovation entirely.

This book provides frameworks for understanding and addressing the ethical, organizational, and societal implications of AI-assisted decision-making. It examines how AI systems currently function in strategic contexts, what changes when algorithms influence critical choices, and how organizations can capture competitive advantages while maintaining human accountability. The exploration draws on real-world implementations across industries, emerging governance models from leading organizations, and practical insights from executives, policymakers, technologists, and ethicists grappling with these challenges.

The approach throughout emphasizes actionable understanding over abstract theorizing. Readers will encounter concrete scenarios that illuminate ethical dilemmas, assessment frameworks for evaluating AI readiness, governance structures that assign clear accountability, and implementation strategies that preserve human agency. The goal is not to provide definitive answers to questions that remain contested and context-dependent, but rather to equip professionals with the conceptual tools and practical methodologies needed to make informed decisions about AI adoption in their specific organizational and societal contexts.

The transformation underway requires neither uncritical embrace of automation nor reflexive resistance to technological change. It demands thoughtful governance, cross-disciplinary collaboration, and sustained commitment to preserving human values while leveraging AI capabilities. The following chapters provide the foundation for navigating this transition responsibly, recognizing that the future of AI decision-making depends less on the algorithms themselves than on the human systems of accountability, oversight, and shared purpose built around them.

CHAPTER 1.

THE NEW DECISION-MAKERS: UNDERSTANDING AI IN THE BOARDROOM AND BEYOND

When a major insurance company's AI system began denying claims at rates 30% higher than human adjusters, executives faced an uncomfortable question: Who was actually making the decisions? The algorithm had been trained on historical data, approved by the board, and integrated into workflows, yet no single person could explain why specific claims were rejected. This created an accountability vacuum that would eventually cost the company millions in settlements and irreparable damage to its reputation.

This scenario represents neither an isolated incident nor a cautionary tale from some distant future. It reflects the reality facing organizations across industries today as artificial intelligence systems transition from operational tools to strategic decision-makers influencing outcomes that affect millions of lives. The shift is profound yet often invisible, occurring incrementally as algorithms graduate from automating routine tasks to shaping resource allocations, strategic priorities, and critical choices that once resided firmly within human judgment.

The transformation challenges fundamental assumptions about organizational decision-making. For decades, business leaders operated within established frameworks where accountability followed clear lines—executives made strategic choices, managers implemented them, and individuals bore responsibility for outcomes. AI disrupts this clarity. When an algorithm recommends rejecting a loan application, denying medical treatment, or restructuring operations in ways that eliminate jobs, who holds accountability? The data scientists who built the model? The executives who approved its deployment? The algorithm itself? The question reveals an uncomfortable truth that our existing frameworks for responsibility and oversight were designed for a world where humans made decisions, not one where machines increasingly shape them.

Yet this chapter does not approach AI decision-making as a threat to be feared or a revolution to be resisted. The technology offers genuine advantages—pattern recognition capabilities that exceed human analytical capacity, processing speeds that enable real-time responses to market shifts, and consistency that eliminates certain forms of human bias. Organizations that successfully integrate AI into strategic decision-making gain competitive advantages that cannot be ignored. The challenge lies not in whether to adopt these systems, but in how to govern their influence while preserving human agency and accountability.

Understanding this landscape requires examining AI neither through the lens of technological determinism nor ethical absolutism, but as a capability demanding thoughtful integration into existing organizational structures. The insurance company's accountability vacuum emerged not because the technology was inherently flawed, but because the organization failed to establish governance frameworks before deployment. No one had defined clear responsibility for algorithmic decisions. No mechanisms existed for human oversight when the system's recommendations contradicted human judgment. No processes ensured that the algorithm's pattern detection aligned with the company's values and regulatory obligations.

This chapter establishes the foundation for understanding how AI systems currently function as decision-makers in organizational contexts, who shapes their development and deployment, and what fundamentally changes when machines influence critical choices. It examines the stakeholder landscape—from corporate executives seeking competitive advantage to policymakers grappling with governance frameworks, and from data scientists building the systems to employees whose work these algorithms reshape. Most importantly, it reframes the central question facing professionals across industries: not whether AI should participate in organizational decision-making, but how we design systems, structures, and governance mechanisms that preserve human accountability while leveraging algorithmic capabilities.

The answers matter urgently. AI decision-making systems are already operating in boardrooms, hospitals, financial institutions, and government agencies. The question is whether we will govern their influence thoughtfully or discover accountability gaps only after costly failures reveal what we should have built from the beginning.

From Automation to Autonomy: Tracing AI's Evolution in Organizational Decision-Making

The trajectory from automation to autonomy in organizational decision-making did not occur overnight, nor did it follow a linear path. Understanding this evolution requires examining three distinct phases that transformed how organizations approach algorithmic systems and how each phase expanded the scope of machine influence while simultaneously revealing new governance challenges.

The first phase, spanning roughly from the 1980s through the early 2000s, centered on rule-based automation. Organizations deployed systems that executed predetermined logic. If specific conditions existed, then specific actions followed. Payroll processing exemplified this approach—calculating wages, deductions, and tax withholdings according to explicit formulas programmed by humans. These systems operated within narrow, well-defined domains where every contingency could theoretically be anticipated and coded. The value proposition was straightforward: machines performed repetitive calculations faster and more consistently than humans, thereby reducing errors and freeing employees for higher-value work.

Accountability in this phase remained clear. When a payroll system miscalculated wages, the error traced back to either incorrect programming or faulty input data—both ultimately human responsibilities. The machine executed instructions and humans bore responsibility for those instructions and their consequences. Organizations understood these systems as sophisticated tools, not decision-makers.

The second phase emerged with the machine learning revolution beginning in the mid-2000s and accelerating through the 2010s.[1] Rather than following explicit rules, these systems identified patterns in historical data and made predictions based on those patterns. A credit scoring algorithm, for instance, analyzed thousands of variables across millions of past loan applications to predict default probability for new applicants. Crucially, the system determined which variables mattered and how they related to outcomes—relationships that humans neither specified nor fully understood.

This represented a qualitative shift in capability and complexity. Machine learning systems could handle problems with multiple interacting variables, non-linear relationships, and probabilistic outcomes—precisely the conditions characterizing most real-world organizational decisions.[1] Retailers used these systems to optimize pricing and inventory. Healthcare organizations deployed them to predict patient readmission risks. Financial institutions relied on them for fraud detection and algorithmic trading.

Yet machine learning introduced the transparency problem. As these systems grew more sophisticated, they became less explainable. The algorithms could predict outcomes with impressive accuracy, but even their creators could not always articulate precisely why a particular prediction was made. This opacity created new accountability challenges, particularly in high-stakes domains like hiring, lending, and criminal sentencing where algorithmic recommendations affected fundamental life opportunities.

The third and current phase involves autonomous decision systems that not only analyze and recommend but execute decisions with minimal human intervention. These systems operate in real-time and respond to changing conditions faster than human oversight allows. Algorithmic trading systems execute thousands of transactions per second based on market patterns. Dynamic pricing algorithms adjust prices continuously across millions of products. AI-powered content moderation systems remove posts from social media platforms before human reviewers even see them.

Research from MIT's Initiative on the Digital Economy demonstrates that organizations adopting these autonomous systems report significant efficiency gains but also encounter governance challenges their existing frameworks cannot address. When decisions occur at machine speed, traditional approval processes become bottlenecks. When algorithms optimize for measurable metrics, they may inadvertently undermine unmeasured values. When systems learn continuously from new data, their behavior evolves beyond their original design specifications.

This evolution from automation to autonomy fundamentally transforms the accountability question. Organizations no longer simply ask whether their systems execute instructions correctly, but whether the systems themselves are making appropriate decisions—a question requiring frameworks that most organizations have yet to develop.

The Boardroom Reality: How Executives Currently Use AI for Strategic Choices

In boardrooms across industries, artificial intelligence has quietly transitioned from a technology topic relegated to IT presentations to a regular presence in strategic discussions. Executives now routinely encounter AI-generated insights when making decisions about market entry, capital allocation, product portfolios, and organizational restructuring.[2 8] Yet the reality of how leaders actually use these systems reveals a more nuanced picture than either enthusiastic adoption narratives or cautionary tales suggest.

A 2025 survey of 876 top executives found that nearly half of all AI-related decisions now flow through C-suite leaders, with ceos and chief technology officers holding the largest share of decision-making authority.[6] This concentration reflects a fundamental shift. AI is no longer managed as a back-office technology tool but governed as a strategic transformation lever requiring board-level oversight.[6] The same research revealed that C-level executives collectively account for over three-quarters of AI decisions, encompassing everything from vendor selection and use-case prioritization to policies on risk, ethics, and regulatory compliance.[6]

The domains where executives deploy AI for strategic choices cluster around several recurring themes. In strategy and market positioning, machine learning models simulate market-entry scenarios, estimate competitive responses, and analyze product performance across segments—surfacing non-obvious correlations that challenge conventional assumptions about growth opportunities.[2 5 7 8] For financial planning, AI-driven forecasting systems generate dynamic revenue and cash flow projections that replace static quarterly estimates, while scenario-planning tools support "what-if" analysis for interest rate changes, commodity shocks, or regulatory shifts.[5 7 8] One analysis noted that approximately seventy percent of leading companies report faster executive decisions when using AI tools, a speed advantage that fundamentally alters the cadence of strategic deliberation.[5]

Risk management represents another critical application area. Predictive models provide early-warning indicators for credit exposure, fraud, cyber incidents, and operational failures, while compliance monitoring systems scan transactions and communications for potential violations.[2 5] Boards increasingly review AI-generated risk dashboards alongside traditional risk registers, changing both the granularity and frequency of oversight. In customer and growth strategy, executives use AI to model behavior shifts, churn risk, and lifetime value—insights that inform pricing strategies, retention programs, and decisions about whether to pursue personalization at scale despite privacy concerns.[2 4 7 8]

Yet this widespread adoption coexists with a striking maturity gap. While surveys indicate that seventy-eight percent of companies use AI in at least one business function, research suggests that only approximately one percent have achieved mature, scaled deployment characterized by wide integration, strong governance, and measurable impact.[3 7] Most organizations use AI opportunistically to support isolated decisions rather than as a systematic decision architecture embedded throughout strategic processes.[3]

This gap creates vulnerabilities. Executives may over-trust opaque models without understanding the training data limitations or potential biases, or conversely under-trust useful systems due to insufficient literacy, leading to parallel traditional analyses that waste AI's potential.[3 8 9] Many boards lack clear AI performance indicators, making it difficult to evaluate whether AI-enabled decisions actually improve outcomes.[3, 4] The accountability question becomes particularly acute when AI-driven choices harm stakeholders through discriminatory outcomes, unsafe products, or misleading content, yet no clear escalation path or responsibility framework exists.[4, 8]

The emerging best practice response involves treating AI governance as a board mandate comparable to cybersecurity or audit oversight.[4, 8] Leading organizations establish cross-functional AI steering committees, define ethical principles for acceptable uses, and implement continuous monitoring protocols.[4, 8] They invest in AI literacy programs for senior leaders and align organizational structures—sometimes creating Chief AI Officer roles—with strategic ambitions.[3, 4, 9] These measures aim to ensure that AI insights are not merely consumed but understood, questioned, and used within clearly defined moral and legal boundaries.[4, 8]

The current boardroom reality thus reflects a transitional moment in which AI has become indispensable to strategic decision-making, yet the governance frameworks required to manage its influence responsibly remain incomplete in most organizations.

The Accountability Gap: When Algorithms Make Decisions, Nobody Can Explain

The insurance company scenario that opened this chapter illustrates a phenomenon that philosophers and governance experts now call the accountability gap—situations where algorithmic systems significantly shape or make decisions, yet no human or institution can be clearly identified as responsible in a way that supports justification, explanation, or liability.[11, 12, 13] This gap emerges not from a single technical flaw but from the convergence of three distinct challenges:

algorithmic opacity, diffused responsibility across multiple actors, and the erosion of meaningful human decision ownership.

Algorithmic opacity represents the most visible dimension of the problem. Modern machine learning systems, particularly deep neural networks, operate through millions of weighted connections that identify patterns in training data. These systems can predict loan default risk, medical treatment outcomes, or employee performance with impressive accuracy, yet even their creators often cannot articulate precisely why a specific prediction was generated. The model has learned correlations that may be statistically valid but remain conceptually opaque—a phenomenon researchers describe as the "black box" problem.[11] When a hiring algorithm ranks one candidate above another, the system may be responding to hundreds of interacting variables in ways that defy simple explanation, leaving hiring managers unable to justify their decisions beyond citing the algorithm's recommendation.

Yet opacity alone does not create accountability gaps. Rule-based systems with transparent logic can also generate responsibility vacuums when decision-making authority becomes diffused across designers, deployers, and users.[11] Research on organizational AI adoption reveals a troubling pattern: developers point to users, arguing they only built tools that others chose to deploy, while users point to the algorithm, claiming they followed system recommendations and executives point to vendors, noting they purchased certified solutions.[13] This diffusion creates what ethics scholars identify as a structural accountability gap where harmful outcomes occur but no actor satisfies the conditions for clear responsibility.[13]

The third and perhaps most insidious challenge involves what philosopher Anna Lukina terms attributability gaps—situations where decisions reflect value judgments that no identifiable human has consciously endorsed.[10] Machine learning models embody normative choices at multiple stages: which outcomes to optimize, which trade-offs between competing values to accept, and which historical patterns to treat as legitimate guides for future decisions. When professionals adopt AI outputs without understanding or endorsing these embedded value judgments, decisions cease to be clearly theirs in a morally meaningful sense.[10] A judge using a recidivism prediction algorithm may nominally make the sentencing decision, yet if the algorithm's risk assessment reflects values about rehabilitation versus punishment that the judge has never examined, the decision's normative content has been outsourced in ways that undermine genuine ownership.[10]

Recent research suggests that efforts to address opacity through explainable AI may paradoxically deepen accountability gaps rather than close them.[11] Studies indicate that when algorithms provide explanations for their recommendations, users may perceive the systems as intentional agents worthy of blame, deflecting responsibility away from human decision-makers.[11] Simultaneously, explanations can create false confidence that users understand and control systems whose actual operation remains opaque, leading to over-trust and inappropriate delegation of judgment.[11] Most critically, post-hoc explanations may not empower affected individuals such as job applicants, loan seekers, and patients with meaningful ability to contest decisions, even while legitimizing the system as "transparent."[11]

The accountability gap thus represents not merely a technical challenge requiring better explainability tools, but a fundamental governance problem demanding new frameworks for assigning responsibility, enabling meaningful human oversight, and ensuring that algorithmic influence remains answerable to human values and institutional accountability structures.

Beyond Efficiency: Understanding What Changes When Machines Influence Critical Choices

The efficiency narrative dominates most discussions about AI adoption in organizational decision-making. Executives cite faster processing times, reduced operational costs, and the ability to analyze datasets beyond human capacity. Yet focusing exclusively on efficiency metrics obscures more fundamental transformations that occur when machines begin influencing critical choices—changes that affect not just how quickly decisions are made, but what those decisions mean and who bears responsibility for their consequences.

When AI systems move from supporting decisions to shaping them, the nature of organizational choice-making undergoes several qualitative shifts that efficiency measurements cannot capture. The first involves what decision science researchers call decision ownership, which is the relationship between the person nominally responsible for a choice and the reasoning that produced it. In traditional decision-making, executives may consult advisors, review data, and consider recommendations, but they ultimately synthesize inputs through their own judgment, applying contextual knowledge and value priorities that reflect their understanding of organizational mission and stakeholder obligations.

AI-influenced decisions disrupt this synthesis. When an algorithm recommends a specific course of action based on pattern recognition across millions of data points, the executive faces a fundamentally different cognitive task. Rather than integrating diverse inputs into a coherent judgment, they must decide whether to accept or override a recommendation whose internal logic may be opaque. Research on human-machine decision-making reveals that this shift often transforms decision-makers from active synthesizers into passive validators, particularly when algorithmic recommendations arrive with high confidence scores or when organizational culture treats AI outputs as objective truth rather than probabilistic predictions.

The second transformation concerns value encoding and trade-offs. Every consequential decision involves balancing competing priorities—short-term profit versus long-term sustainability, efficiency versus equity, innovation versus risk management. Human decision-makers navigate these trade-offs through deliberation that weighs organizational values, stakeholder interests, and ethical considerations alongside quantitative metrics. AI systems, by contrast, optimize for specified objectives encoded in their design. When a machine learning model recommends workforce restructuring to maximize quarterly earnings, it reflects the objective function its creators specified, not a holistic assessment of what the organization values.

This distinction matters profoundly in high-stakes contexts. A hiring algorithm optimized for retention rates may systematically disadvantage candidates from non-traditional backgrounds who bring valuable diversity but don't match historical patterns. A resource allocation system optimized for utilization efficiency may recommend closing facilities that serve marginalized communities, treating social mission as external rather than core value. The machine has not failed technically—it has optimized precisely as designed. Yet the decision it produces may violate values that no one explicitly chose to subordinate.

The third shift involves temporal dynamics and reversibility. Human decision-makers typically approach critical choices with awareness of their provisional nature. They understand that circumstances change, new information emerges, and course corrections may be necessary. AI systems operating at scale and speed can execute decisions across thousands of cases before problems become apparent. When an insurance claims algorithm systematically denies coverage to a demographic group, the harm accumulates across numerous individual decisions before patterns trigger human review. The efficiency that makes AI valuable—processing volume and speed—simultaneously amplifies the consequences of flawed decision logic.

These transformations reveal why governance frameworks focused solely on algorithmic accuracy or processing efficiency prove inadequate. The question is not whether AI makes decisions faster or handles more data, but whether organizations maintain meaningful human agency over the values embedded in those decisions, the trade-offs they reflect, and the accountability structures that govern their consequences. Efficiency gains matter, but they represent only the most visible dimension of a far more complex transformation in how organizations make choices that affect stakeholders' lives and opportunities.

The Stakeholder Landscape: Who Shapes AI Decision-Making Systems and Why It Matters

Understanding who shapes AI decision-making systems requires mapping a complex ecosystem of actors whose choices, incentives, and power dynamics determine how these technologies are designed, deployed, and governed. The stakeholder landscape extends far beyond the data scientists who build algorithms and encompasses corporate leaders who authorize deployment, domain experts whose judgment AI systems augment or replace, regulators who establish guardrails, and communities whose lives these systems affect.[14, 15] Each group wields distinct forms of influence—technical, economic, legal, or social—and their interactions determine whether AI decision-makers serve broad societal interests or narrow organizational objectives.[14, 15]

Corporate executives and boards occupy the most visible position in this landscape, wielding formal authority over AI adoption decisions. Research from 2025 examining 876 top executives found that nearly half of all AI-related decisions now flow through C-suite leaders, with ceos and chief technology officers holding the largest share of decision-making authority. This concentration reflects AI's elevation from back-office technology to strategic transformation lever requiring board-level oversight. These leaders define acceptable risk levels, allocate resources for model validation and ethics teams, and establish governance structures that integrate AI into enterprise risk management frameworks.[15] Their choices about performance metrics directly shape model objectives—whether systems optimize short-term revenue, balance compliance and reputational risk, or incorporate social impact alongside profit.[15]

AI researchers, data scientists, and engineers translate abstract business goals into mathematical formats, wielding what scholars call epistemic authority or the power to define how problems are represented and solved. They formalize decision problems using frameworks from probability theory and utility theory, choosing algorithms and objective functions that encode which errors or harms are tolerated.[14] As decision theorist Mykel Kochenderfer notes in his work on algorithms under uncertainty, these systems operationalize the maximum expected utility principle from economics, meaning technical teams effectively decide whose utility matters and how competing values are weighted.[14] When these professionals lack domain expertise or stakeholder input, they may implicitly encode biases or narrow objectives into AI behavior that later proves difficult to detect or reverse.[14]

Domain experts and frontline professionals like clinicians, underwriters, social workers, and HR specialists occupy a critical but often marginalized position. They provide the contextual knowledge that should inform feature selection and causal assumptions, yet are frequently excluded from design processes.[14] How these practitioners trust, override, or circumvent AI recommendations shapes real-world outcomes regardless of model predictions.[15] Professional codes imposing standards for consent, explanation, and due process can constrain AI deployment, but only when domain experts retain sufficient authority to enforce these norms against efficiency pressures.

Affected communities and data subjects represent those most vulnerable to algorithmic harm yet typically possess the least formal power in system design. Legal frameworks grant individuals some rights over data collection and processing, while public pressure through boycotts and social movements can force organizations to withdraw harmful systems. Some organizations experiment with participatory design approaches—co-design workshops, citizen juries, community advisory boards—to incorporate affected perspectives into problem framing and acceptable use definitions. Yet these mechanisms remain an exception rather than standard practice.

Regulators and policymakers articulate societal guardrails through binding rules on data protection, discrimination, safety, and accountability.[15] Their design choices—risk-based regulation, audit obligations, rights to explanation—determine baseline protections for citizens.[15] Increasingly, governments themselves deploy AI for public opinion analysis and resource allocation, setting precedents for acceptable practices while grappling with the same governance challenges facing private organizations.[15]

This stakeholder landscape matters profoundly because AI decision-making systems operationalize specific theories of rational choice and utility maximization.[14] The architecture of participation determines whose values are encoded into these systems, whose interests are protected or sacrificed, and whether accountability structures can prevent responsibility diffusion when harms occur.[14] [15] Building legitimate AI decision-makers requires not just technical excellence but inclusive governance that balances power across this diverse ecosystem.

Reframing the Question: Not Whether to Use AI, But How to Govern Its Influence

The debate over AI adoption has largely concluded in most sectors. Research from 2025 indicates that approximately seventy-five percent of companies already use generative AI in some capacity by integrating these systems into content creation, customer service, software development, and strategic analysis. The question facing organizations is no longer whether to deploy AI decision-making systems, but how to establish governance frameworks that constrain their influence while capturing their capabilities.[18 19] This reframing shifts the conversation from technology procurement to institutional design regarding the structures, norms, and accountability mechanisms that determine what AI systems are permitted to do, how they are overseen, and who bears responsibility when they cause harm.

This transition from adoption debate to governance imperative reflects a fundamental recognition: AI systems now function as a distinct class of organizational actor that shapes outcomes even when humans nominally retain final authority.[16 17] Machine learning models triage job applicants, prioritize patients for treatment, recommend credit limits, adjust prices dynamically, and increasingly advise executives on strategic choices.[16 17] Even when formally classified as decision support tools, these systems exert influence by directing human attention toward specific options, constraining the range of alternatives considered, and shifting the burden of proof onto those who would deviate from algorithmic recommendations. The relevant question is not whether AI "decides" in some metaphysical sense, but how much causal and normative influence it exercises over outcomes, and under what conditions that influence remains legitimate and controllable.

Governing AI influence requires treating models, data, and organizational practices as an interlocking sociotechnical system rather than isolated technical tools. Data choices embed value-laden judgments about what constitutes success and whose experiences matter. Model design encodes trade-offs between competing priorities—accuracy versus fairness, efficiency versus safety, short-term profit versus long-term sustainability. Deployment contexts create feedback loops that can amplify initial biases or generate emergent behaviors that designers never anticipated. Effective governance must therefore align technical controls with organizational purpose, risk tolerance, and external obligations, thereby integrating oversight mechanisms throughout the AI lifecycle rather than treating governance as a one-time approval gate.[17 18]

Contemporary AI governance frameworks typically encompass several interconnected domains.[17 18 19] Strategy and purpose articulate why AI is used and what outcomes it should or must not pursue.[18 19] Risk management identifies, assesses, and mitigates potential harms including bias, privacy violations, security vulnerabilities, and systemic instability.[17 18] Ethics and human rights considerations embed principles like fairness, accountability, and transparency into concrete requirements and controls.[16 17] Compliance mapping connects AI use cases to evolving legal regimes, from the European Union's risk-based AI Act to sector-specific regulations in healthcare, finance, and employment.[19 20] Operational controls govern data quality, model validation, performance monitoring, incident response, and system decommissioning.[16 17] Organizational structures establish councils, assign accountable ownership, and create escalation paths for ethical concerns.[16 18]

The gap between AI adoption and governance implementation poses significant risks. The same research showing seventy-five percent AI adoption found that only about one-third of organizations had implemented responsible controls defined to include governance policies, risk assessments, and oversight mechanisms. This mismatch between rapid capability diffusion and lagging governance structures creates vulnerabilities such as data leaks exposing confidential information, biased outputs that damage trust and invite legal scrutiny, uncontrolled experimentation driving unsustainable costs, and regulatory breaches as legal frameworks tighten.[17] These failures represent not merely operational mishaps but fundamental breakdowns in how organizations constrain AI's power to reshape opportunity, allocate resources, and distribute harm at scale.

Reframing the AI question from adoption to governance transforms how leaders approach these systems by shifting focus from acquiring capabilities to building the moral and institutional infrastructure that determines how those capabilities serve human values and organizational accountability.[18] [19]The insurance company's accountability vacuum described at this chapter's opening represents neither an anomaly nor a worst-case scenario, but rather a preview of challenges that organizations across industries will increasingly confront as AI systems transition from operational tools to strategic decision-makers. The fundamental question is not whether these systems will influence critical choices (that transformation is already underway) but whether organizations will establish governance frameworks before accountability gaps create consequences that could have been prevented.

This chapter has traced AI's evolution from rule-based automation through machine learning to autonomous decision systems, revealing how each phase expanded algorithmic influence while simultaneously eroding traditional accountability structures. The progression matters because it demonstrates that current governance challenges emerged gradually, through incremental choices about system design, deployment contexts, and oversight mechanisms. Organizations did not suddenly lose control of AI decision-making; they systematically delegated authority without building corresponding accountability architectures, creating the responsibility vacuums that now demand urgent attention.

The boardroom reality examined here reveals a striking paradox. Executives increasingly rely on AI for strategic insights while most organizations lack the governance maturity required to manage these systems responsibly. Nearly half of all AI-related decisions now flow through C-suite leaders, yet only approximately one percent of companies have achieved scaled deployment with strong governance and measurable impact. This gap between adoption enthusiasm and institutional readiness creates vulnerabilities that extend far beyond operational inefficiencies, threatening stakeholder trust, regulatory compliance, and the legitimacy of algorithmic influence itself.

Understanding what changes when machines influence critical choices requires looking beyond efficiency metrics to examine transformations in decision ownership, value encoding, and accountability structures. AI systems do not simply accelerate existing decision processes; they fundamentally alter the relationship between decision-makers and the reasoning that produces choices. When executives validate algorithmic recommendations rather than synthesizing diverse inputs through their own judgment, when optimization objectives encode value trade-offs that no human has consciously endorsed, when decisions execute at scales and speeds that prevent meaningful oversight, these shifts demand governance frameworks that most organizations have yet to develop.

The stakeholder landscape mapping reveals why technical solutions alone cannot address these challenges. AI decision-making systems reflect choices made by corporate leaders, data scientists, domain experts, regulators, and affected communities—each wielding distinct forms of influence over how these technologies are designed, deployed, and constrained. Building legitimate AI decision-makers requires not just algorithmic accuracy but inclusive governance that balances power across this diverse ecosystem, ensuring that the values embedded in these systems reflect broader societal interests rather than narrow organizational objectives.

Reframing the central question from whether to adopt AI toward how to govern its influence represents the critical shift required for responsible integration. The adoption debate has concluded, and organizations now face the harder work of building moral and institutional infrastructure that determines how algorithmic capabilities serve human values and organizational accountability. This infrastructure encompasses strategy articulation, risk management protocols, ethics principles, compliance frameworks, operational controls, and organizational structures that assign clear responsibility for AI decisions and their consequences.

The chapters that follow build on this foundation by examining how organizations can translate governance principles into operational reality—from understanding AI's analytical capabilities and limitations to implementing accountability frameworks, designing human-machine collaboration models, and ultimately creating shared ethical standards for an AI-augmented world. The insurance company's costly failure provides the cautionary tale, while the task ahead is building the frameworks that prevent such failures from becoming the norm rather than the exception.

CHAPTER 2.

AUTOMATED BUSINESS INTELLIGENCE: WHEN ALGORITHMS KNOW MORE THAN WE DO

A global retailer's pricing algorithm detected a pattern that contradicted twenty years of industry wisdom: customers were more likely to purchase premium products when prices increased during specific weather conditions in particular regions. The human merchandising team initially dismissed the finding as a statistical anomaly, yet when they reluctantly tested the algorithm's recommendation, quarterly revenue in those categories increased by 18%, which forced executives to confront an uncomfortable reality about the limits of human intuition in data-rich environments.

This scenario represents a fundamental shift in how organizations generate and apply business intelligence. For decades, competitive advantage depended on human expertise. Including seasoned analysts who could interpret market signals, executives whose intuition was honed by years of experience, and domain specialists who understood their industries deeply. These human capabilities remain valuable, yet they now operate alongside machine learning systems that process millions of data points simultaneously, identifying correlations and patterns at scales that exceed human cognitive capacity.

The transformation extends beyond simple automation. Early business intelligence systems helped humans analyze data faster, but contemporary AI algorithms don't just accelerate existing analytical processes, they detect relationships that humans wouldn't think to look for, predict outcomes based on variables humans might not consider relevant, and generate insights that contradict conventional wisdom. When an algorithm identifies a profitable strategy that defies industry experience, organizations face a critical question: whose knowledge should guide decisions?

This chapter examines the implications of this knowledge shift across multiple dimensions. The pattern recognition advantage explores how machine learning algorithms identify signals in complex datasets that remain invisible to human analysts, even highly skilled ones. Predictive analytics investigates the transition from reactive decision-making based on historical understanding to anticipatory strategies driven by algorithmic forecasting. The interpretation challenge addresses a fundamental tension: algorithms excel at detecting patterns but often cannot explain why those patterns exist or what causal mechanisms drive them.

The organizational consequences of algorithmic intelligence extend beyond individual decisions. When critical business knowledge increasingly resides within machine learning models rather than human expertise, companies must reconsider how they develop strategy, train employees, and maintain institutional memory. A retail executive who spent twenty years learning customer behavior may find their knowledge contradicted by an algorithm trained on millions of transactions. A supply chain manager's hard-won understanding of logistics patterns may be superseded by a system that optimizes routes using variables the manager never considered.

These scenarios create what we might call the confidence paradox — situations where algorithmic certainty conflicts with human contextual judgment. Machine learning systems generate predictions with statistical confidence levels, yet those confidence metrics don't account for factors the algorithm wasn't trained to recognize. A pricing algorithm might be 95% confident in its recommendation based on historical data, but it cannot assess whether current market conditions represent a fundamental shift that makes historical patterns irrelevant.

The challenge for organizations is not choosing between human intelligence and artificial intelligence but rather establishing governance frameworks that leverage algorithmic pattern detection while preserving human judgment about context, causation, and strategic fit. This requires answering difficult questions: When should executives trust machine insights that contradict their experience? How do organizations validate algorithmic recommendations without undermining the speed advantages that make AI valuable? What happens to professional expertise when algorithms consistently outperform human analysis?

This chapter provides frameworks for navigating these questions as it examines how leading organizations balance algorithmic insights with human interpretation. The goal is not to resolve the tension between machine and human intelligence, but to understand how that tension can drive better decision-making when managed thoughtfully transforming the relationship between data, algorithms, and human judgment in ways that enhance rather than diminish organizational capability.

The Pattern Recognition Advantage: How Machine Learning Identifies What Humans Cannot See

Machine learning's most distinctive capability lies not in processing data faster than humans, but in detecting multidimensional patterns that remain fundamentally invisible to human cognition. While experienced analysts excel at tracking relationships among a handful of variables like price trends, seasonal fluctuations, and customer demographics, contemporary pattern recognition algorithms operate across hundreds or thousands of dimensions simultaneously, identifying correlations that no human could systematically monitor or even conceptualize.

Consider the cognitive limitations inherent in human pattern recognition. Research on decision-making shows that even domain experts rely on patterns matching for approximately ninety percent of their challenging decisions, yet this expertise operates within narrow bounds.[22] A retail merchandiser might intuitively grasp how weather affects sales in familiar product categories by drawing on years of observation. That same expert, however, cannot simultaneously track how weather interacts with hundreds of other variables such as day of week, local events, competitor pricing, social media sentiment, inventory levels at nearby stores, and customer browsing patterns from weeks earlier to predict purchasing behavior across thousands of products.

Machine learning systems face no such constraints. Pattern recognition algorithms can invest millions of historical transactions, each characterized by hundreds of features, and identify which combinations of variables reliably predict outcomes.[21 24] These systems discover what researchers call high-dimensional patterns: regularities that only become apparent when many variables are considered together, where no single factor is decisive but the combination proves predictive.

The manufacturing sector provides concrete evidence of this advantage. Sensor networks in industrial facilities generate continuous streams of data—temperatures, vibrations, pressures, flow rates—from equipment throughout production lines. Human engineers monitoring these systems can track individual sensors and respond to obvious anomalies, but they cannot simultaneously process patterns across dozens of sensors over extended time periods. Pattern recognition models trained on historical sensor data have successfully identified subtle combinations of readings that precede equipment failures days or weeks in advance, enabling predictive maintenance that reduces unplanned downtime. These failure signatures often involve small deviations across multiple sensors that individually appear within normal ranges and are patterns that remain invisible to human operators precisely because they require integrating information beyond human working memory capacity.

The healthcare domain demonstrates similar capabilities in medical imaging. Radiologists possess remarkable visual pattern recognition skills honed through years of training, yet they examine images sequentially and can experience fatigue or inconsistency. Machine learning systems trained on millions of labeled medical images can detect subtle texture patterns, shape irregularities, or pixel-level variations associated with early-stage diseases.[24] Studies have documented cases where algorithms identified pathological features that human radiologists initially missed—not because the humans lacked expertise, but because the relevant patterns existed at scales or involved combinations of features that human visual processing does not naturally integrate.

This pattern recognition advantage extends to temporal dimensions as well. Humans exhibit recency bias and struggle to maintain detailed memory of long, complex sequences. Machine learning models can analyze complete interaction histories—every customer service contact, website visit, purchase, and support ticket over years—to identify behavioral trajectories that predict outcomes like customer churn or fraud. These temporal patterns often involve subtle shifts in engagement frequency, channel preferences, or interaction timing that accumulate meaning only over extended periods.

The business intelligence implications are profound. Organizations now possess systems that can surface opportunities and risks embedded in operational data that would remain hidden to human analysis regardless of expertise or effort invested.[23 24] A pricing algorithm might discover that customer willingness to pay for certain products increases under specific combinations of inventory levels, competitive positioning, and external economic indicators—a multidimensional relationship no human analyst would hypothesize to test. A fraud detection system might flag transaction sequences where each individual transaction appears legitimate but the pattern across accounts, timing, amounts, and merchant categories indicates coordinated criminal activity.

Yet this capability creates fundamental tension. When algorithms consistently detect patterns that contradict human intuition or reveal relationships humans cannot independently verify, organizations must decide whose knowledge should guide decisions—and who bears responsibility when those decisions cause harm.

From Hindsight to Foresight: Predictive Analytics and the Shift from Reactive to Anticipatory Decision-Making

Traditional business intelligence has operated primarily in the past tense, answering questions about what happened and why.[28] Quarterly reports dissect previous performance, dashboards visualize historical trends, and post-mortem analyses explain missed targets or unexpected successes. This retrospective orientation served organizations well when competitive cycles moved slowly and market conditions changed gradually, which allowed leaders to extrapolate from past patterns with reasonable confidence. Predictive analytics fundamentally alters this temporal relationship, shifting organizational attention from hindsight to foresight—from understanding what occurred to anticipating what will happen next.[28]

This transition represents more than incremental improvement in forecasting accuracy. Predictive analytics enables what researchers call anticipatory decision-making, which is organizational systems that continuously monitor signals, infer emerging futures, and trigger responses before events fully materialize.[27] [26] A financial institution using traditional analytics might analyze loan default patterns quarterly to adjust risk policies retrospectively. That same institution using predictive models can score every application in real time, flagging elevated risk profiles before defaults occur and adjusting lending criteria dynamically as economic indicators shift. The decision architecture moves from periodic review cycles to continuous sensing and response.[27]

The mechanisms underlying this shift rest on statistical modeling techniques that estimate future outcomes based on historical and current data.[29 28] Regression models identify relationships between variables like how pricing affects demand and how customer behaviors predict churn which then enables organizations to forecast outcomes under different scenarios.[25 29] Decision trees segment populations into risk categories or opportunity clusters, translating complex datasets into actionable classifications.[25 29] Neural networks and machine learning approaches detect nonlinear patterns in high-dimensional data, often outperforming simpler methods in volatile environments where relationships between variables change over time.[27]

Research on strategic foresight demonstrates measurable performance advantages for organizations that successfully embed predictive capabilities into decision workflows. Studies document that firms integrating predictive analytics into strategic planning achieve faster market response times and higher profitability from strategic initiatives compared to those relying primarily on retrospective analysis.[27] Government agencies using predictive models to identify infrastructure at risk of failure can schedule preventive maintenance, reducing both catastrophic failures and long-term costs compared to reactive repair strategies.[30] These gains emerge not simply from better forecasts, but from decision translation—the systematic incorporation of predictions into operational triggers, resource allocation rules, and strategic commitments.[27 28]

Yet this temporal reorientation creates new governance challenges. When organizations act on predicted futures rather than observed realities, the basis for accountability shifts. A manager who responds to an actual customer complaint operates in a framework of clear cause and effect. A manager who preemptively contacts customers flagged by a churn prediction model acts on probabilistic inference. Some of the flagged customers would have stayed regardless, while others not flagged may leave unexpectedly. The algorithm's confidence level provides statistical justification, but it cannot account for contextual factors the model wasn't trained to recognize such as a competitor's imminent product launch, regulatory changes, or shifting customer values that historical data doesn't capture.

This creates what might be termed the foresight accountability gap. Organizations must determine who bears responsibility when anticipatory decisions based on algorithmic predictions produce unintended consequences. If a predictive maintenance system recommends deferring equipment service and a failure occurs, is the maintenance team accountable for following the algorithm's guidance, or are data scientists responsible for model limitations? If a demand forecasting model drives inventory decisions that result in stockouts during an unprecedented market shift, where does accountability reside?

The shift from reactive to anticipatory decision-making thus requires not only technical capabilities but governance frameworks that address how organizations validate predictions, override algorithmic recommendations when contextual judgment demands it, and assign responsibility when foresight proves imperfect.[27 30] The competitive advantages of predictive analytics are substantial, yet capturing those benefits responsibly demands careful attention to the human systems of oversight and accountability that surround algorithmic foresight.[27 28]

The Interpretation Challenge: When Algorithms Detect Patterns Without Explaining Causation

The most profound limitation of pattern-detecting algorithms lies not in what they find, but in what they cannot explain.[1 32] When a machine learning system identifies that customers with a specific combination of browsing behaviors, purchase timing, and demographic characteristics are 73% more likely to churn, it provides a statistically robust prediction without offering any account of why that relationship exists. This gap between detection and explanation creates fundamental challenges for organizations that must justify decisions, assign accountability, and ensure that automated intelligence serves rather than subverts human values.[1 32]

The distinction between correlation and causation becomes operationally critical when algorithms drive business decisions. Traditional statistical analysis has always warned against confusing association with causation, yet human analysts typically develop hypotheses about underlying mechanisms like theories about customer motivation, market dynamics, or operational constraints that explain observed patterns. Machine learning systems, by contrast, optimize purely for predictive accuracy.[1 32] They exploit whatever correlations exist in training data, regardless of whether those correlations reflect genuine causal relationships or spurious associations that happen to hold in historical datasets.[1 32]

Research on algorithmic decision-making demonstrates that this limitation has measurable consequences across industries. In credit scoring, models may incorporate variables that correlate with loan default without capturing the actual drivers of financial distress.[31] A postal code might predict repayment rates because it serves as a proxy for socioeconomic factors, employment stability, or access to financial services, yet the model treats geography as predictive without understanding these underlying mechanisms. When such systems deny credit, institutions struggle to provide meaningful explanations beyond "your application scored below our threshold," which fails to satisfy both regulatory requirements for adverse action notices and basic standards of procedural fairness.[31]

The interpretation challenge intensifies when algorithmic recommendations contradict domain expertise. A pricing algorithm might detect that revenue increases when certain products are discounted during specific weather conditions in particular regions—a multidimensional pattern no human analyst would hypothesize. Executives face a dilemma: the algorithm's track record demonstrates predictive success, yet no one can articulate a causal mechanism that would explain why the pattern holds. Acting on unexplained correlations may capture short-term gains while exposing the organization to risks the model cannot anticipate including market shifts, competitor responses, or customer perception changes that fall outside the training data's scope.

Healthcare applications illustrate how pattern detection without causal understanding affects high-stakes decisions. Diagnostic algorithms trained on millions of medical images can identify subtle texture patterns or pixel-level variations associated with disease, sometimes detecting pathologies that radiologists initially miss.[31] These systems achieve their performance through statistical pattern matching rather than understanding disease mechanisms, cellular biology, or progression pathways.[31] When patients or clinicians ask why a scan was flagged as concerning, or whether lifestyle modifications might alter risk, the algorithm cannot answer. It knows what patterns correlate with disease, but not why those patterns emerge or what interventions might disrupt causal chains.

The governance implications extend beyond individual decisions to organizational knowledge and accountability structures. When critical business intelligence resides in algorithmic systems that detect patterns without explaining them, companies must determine who bears responsibility for outcomes. If a fraud detection system blocks legitimate transactions based on correlations in historical fraud data, accountability becomes diffuse—data scientists built a statistically valid model, business leaders approved its deployment, and operations staff followed its recommendations, yet no individual can explain why specific transactions triggered alerts.[31]

This creates what might be termed the explanation deficit: organizations possess systems that reliably predict outcomes yet cannot provide causally grounded justifications for the actions those predictions inform. Addressing this deficit requires governance frameworks that distinguish between contexts where unexplained patterns provide sufficient basis for action and situations where causal understanding remains ethically or operationally necessary—a boundary that depends on stakes, reversibility, and the potential for algorithmic recommendations to encode historical biases as if they were natural laws.[1 32]

Organizational Knowledge Transfer: What Happens When Critical Business Intelligence Resides in Algorithms

When a global pharmaceutical company's drug discovery algorithm identified a promising compound that contradicted the research team's understanding of the disease mechanism, the organization faced an unexpected dilemma. The model had been trained on millions of molecular interactions and clinical trial results, developing pattern recognition capabilities that exceeded any individual scientist's knowledge. Yet when researchers asked the system to explain why this compound showed promise, the algorithm could only point to statistical correlations across thousands of variables. It possessed predictive intelligence without causal understanding. The company's competitive advantage increasingly resided not in its scientists' expertise, but in algorithmic knowledge that those same scientists could neither fully comprehend nor independently replicate.

This scenario illustrates a fundamental transformation in how organizations store and access critical business intelligence. Traditionally, institutional knowledge resided in human experts, documented processes, and organizational routines that could be transferred through mentorship, training programs, and written procedures. Contemporary machine learning systems create a different knowledge architecture wherein expertise becomes encoded in model weights, training data, and algorithmic decision rules that operate beyond direct human comprehension.[33] [1] When a pricing algorithm consistently outperforms experienced merchandisers, or a fraud detection system identifies threats that security analysts miss, the organization's competitive intelligence has effectively migrated from human minds to computational systems.

This migration creates several organizational vulnerabilities that extend beyond technical system failures. Knowledge concentration risk emerges when critical capabilities depend on algorithms that few employees understand. A financial institution may employ hundreds of loan officers who execute decisions, yet only a handful of data scientists who comprehend how the underlying credit scoring models generate recommendations. If those specialists leave or if the vendor relationship ends, the organization faces not just a staffing gap but a knowledge void, because the institutional understanding of why certain lending decisions are made resides primarily in systems rather than people.

Research on organizational learning demonstrates that expertise develops through repeated decision-making cycles where humans receive feedback, adjust their mental models, and refine their judgment over time. When algorithms automate these decision cycles, employees lose opportunities for experiential learning. A customer service representative who follows an AI system's recommendations for resolving complaints may execute tasks efficiently without developing the diagnostic skills to handle novel situations when the algorithm encounters scenarios outside its training data. This phenomenon, which researchers term algorithmic de-skilling, gradually erodes the human capabilities that organizations need when automated systems fail or when market conditions shift beyond historical patterns.

The succession planning implications prove equally significant. Organizations have long managed knowledge transfer through documentation, apprenticeship, and institutional memory preserved by long-tenured employees. When critical intelligence resides in algorithms, succession requires different mechanisms. A retail executive who retires takes decades of merchandising intuition but leaves behind documented strategies and trained successors. A machine learning model that drives pricing decisions across thousands of products cannot be similarly transferred, as its "knowledge" exists as millions of numerical parameters whose collective behavior produces outcomes that may not be reproducible if the model must be rebuilt from scratch.

These challenges demand governance frameworks that treat algorithmic knowledge as an organizational asset requiring active management. Leading organizations implement model documentation standards that record not just technical specifications, but the business logic, assumptions, and constraints embedded in algorithmic systems. They maintain human centers of excellence in critical decision domains even when automation handles routine cases, preserving expertise that can evaluate algorithmic recommendations and intervene when contextual factors demand human judgment. They design hybrid decision architectures where humans and machines contribute complementary capabilities rather than systems that position humans as mere ratifiers of algorithmic outputs they cannot meaningfully assess.[31]

The question facing organizations is not whether to leverage algorithmic intelligence—competitive pressures make that choice increasingly unavoidable—but how to structure knowledge transfer so that critical business intelligence remains accessible, governable, and aligned with human values even when it resides in computational systems.[34]

The Confidence Paradox: Balancing Algorithmic Certainty with Human Contextual Judgment

Machine learning systems generate predictions accompanied by numerical confidence probabilities, risk ratings, certainty percentages—that appear to offer clear guidance for decision-making.[1 32] A credit scoring algorithm assigns an applicant a default probability of 8.3%. A demand forecasting model predicts next quarter's sales with 92% confidence. A fraud detection system flags a transaction with a risk score of 0.87. These quantified assessments create an aura of precision that can be deeply compelling to executives seeking data-driven decisions, yet they mask a fundamental limitation in that statistical confidence measures how well a prediction aligns with historical patterns, not whether acting on that prediction serves organizational values, stakeholder interests, or ethical principles.[34]

This gap between algorithmic certainty and contextual appropriateness creates the confidence paradox.[1 32 34] The more confident an AI system appears (the higher its accuracy metrics, the more decisive its recommendations) the more it can suppress the exercise of human judgment precisely when that judgment matters most.[34] Research on human-machine decision-making demonstrates that people exhibit systematic biases when interpreting algorithmic outputs. They may over-trust systems that display high confidence scores and accept recommendations without adequate scrutiny, or they may under-trust algorithms after witnessing a single visible error and reject superior statistical guidance based on anecdotal experience.[1 31 34]

The paradox intensifies because algorithmic confidence operates in a fundamentally different domain than human contextual judgment.[34] When a pricing algorithm expresses 95% confidence in a recommendation, that number reflects the model's internal consistency with training data or how reliably the identified pattern has held in historical scenarios.[1 32] It cannot account for factors the system was never designed to recognize like an emerging competitor strategy, shifting customer values, regulatory changes on the horizon, or the reputational consequences of being perceived as exploitative even when technically optimizing revenue.[1 34] A human executive reviewing that same pricing decision brings different knowledge to bear such as an understanding of long-term customer relationships, awareness of brand positioning, and sensitivity to fairness concerns that transcend profit maximization.[1 34]

Organizations face the challenge of designing decision architectures that leverage algorithmic pattern detection without allowing numerical confidence to overwhelm human deliberation about purpose and values.[1 31 34] Legal scholars analyzing optimal human-machine collaboration have identified several structural approaches for managing this tension.[31] In human-on-the-loop systems, algorithms generate recommendations but humans must actively approve decisions, particularly for high-stakes, low-volume choices where contextual factors prove decisive.[31] In human-over-the-loop architectures, machines act by default but humans can quickly override automated decisions. This is a model common in fraud detection where speed matters, yet false positives require rapid correction based on contextual knowledge the algorithm lacks.[31]

More sophisticated frameworks recognize that not all human-machine disagreements should be resolved the same way.[31] In medical diagnostics, some institutions implement equality protocols where if either the human clinician or the AI system flags a potential cancer risk, the case proceeds to biopsy, thereby prioritizing the catastrophic cost of false negatives over the inconvenience of additional testing.[31] Conversely, in safety-critical transportation systems, machine authority may override human judgment when sensors detect driver impairment, privileging algorithmic assessment where human reliability is predictably compromised.[31]

The design question is not whether to trust algorithms or humans, but how to structure their interaction so that statistical confidence informs rather than dictates decisions, and human contextual judgment operates as genuine authority rather than perfunctory ratification.[1 31 34] This requires organizations to explicitly map decision types, assign clear override protocols, surface uncertainty and disagreement as valuable signals rather than system failures, and align incentives so that appropriate human resistance to algorithmic recommendations is rewarded rather than penalized.[1 31 34] The confidence paradox cannot be resolved through better algorithms alone but demands governance frameworks that treat the relationship between machine certainty and human judgment as a designed system requiring ongoing calibration and moral accountability.[1 32 34]

Governance Frameworks for Algorithmic Insights: Establishing When to Trust the Machine and When to Override

Establishing when to trust algorithmic recommendations and when human judgment must prevail requires more than good intentions, it demands systematic governance frameworks that translate ethical principles into operational decision rules. Organizations increasingly recognize that the question is not whether to use AI-powered business intelligence, but how to structure oversight mechanisms that preserve human accountability while capturing algorithmic advantages.

The European Union's AI Act provides a regulatory template that many organizations now use as a governance baseline, even beyond EU jurisdictions.[35 37] The legislation introduces a risk-based classification system that explicitly ties oversight requirements to potential impact.[35 37] High-risk applications—including credit scoring, employment decisions, and certain resource allocation systems—must incorporate human oversight mechanisms that enable effective intervention and override capabilities.[35 37] The framework effectively establishes a legal floor that certain decisions cannot be fully automated regardless of algorithmic performance, requiring documented human review with sufficient information to question and reverse machine recommendations.[35 37]

The NIST AI Risk Management Framework offers complementary guidance focused on continuous assessment rather than binary approval processes.[37] Organizations implementing NIST-aligned governance define explicit risk tolerances and impact thresholds that trigger escalation from automated execution to human review. A financial institution might allow algorithmic approval of routine transactions below specified amounts while routing higher-value or anomalous cases to human analysts. The framework emphasizes measurement and monitoring throughout the AI lifecycle, recognizing that systems performing well at deployment may degrade or encounter novel scenarios that demand human intervention.[37]

Translating these external frameworks into internal governance requires organizations to operationalize abstract principles through concrete workflow designs. Leading enterprises implement human-in-the-loop, human-on-the-loop, and human-out-of-the-loop architectures, depending on what is appropriate for different risk profiles. Human-in-the-loop systems require active approval before algorithmic recommendations execute and are common in hiring decisions or insurance underwriting where regulatory requirements or organizational values demand documented human judgment. Human-on-the-loop architecture allows algorithms to operate autonomously under supervision, with humans empowered to intervene when monitoring systems detect anomalies, performance degradation, or ethical concerns. Human-out-of-the-loop automation may be acceptable for low-stakes micro-decisions like email send-time optimization, but even these systems benefit from periodic audit to ensure they haven't drifted toward unintended discriminatory patterns.[35 37]

Effective governance frameworks establish multidimensional criteria that determine when algorithmic insights warrant trust. These criteria typically include validated performance across relevant populations, availability of intelligible explanations that decision-makers can evaluate,[35 36 37] documented data quality and representativeness,[35 36] formal fairness assessments where legally or ethically required,[35 36 37] and demonstrated robustness against adversarial manipulation.[17 36] Organizations codify these dimensions into decision matrices that specify which combinations of model characteristics, decision stakes, and contextual factors permit automation versus requiring human review.

The governance challenge continues to grow because appropriate trust calibration changes over time. Models that perform reliably in stable conditions may fail when market dynamics shift, new competitors emerge, or customer preferences evolve beyond training data patterns. Robust frameworks therefore incorporate continuous monitoring with automated alerts that flag potential problems such as unexpected variance in outcomes across demographic groups, sudden drops in prediction confidence, or patterns suggesting adversarial gaming. These monitoring systems must connect to clear escalation protocols specifying who has authority to suspend automation, what evidence justifies override, and how incidents trigger formal review processes that update governance rules based on lessons learned.[17 36]

The organizational architecture supporting these frameworks proves as critical as the technical systems themselves. Effective governance requires designated model owners accountable for performance and fairness, data stewards responsible for quality and representativeness,[35 36] domain experts empowered to question algorithmic outputs based on contextual knowledge, and executive leadership that sets risk appetite and ensures resources for meaningful oversight rather than perfunctory compliance theater.[35 17 36 37]The transformation from human-centered business intelligence to algorithmic pattern detection represents more than a technological upgrade—it fundamentally alters where organizational knowledge resides, how decisions get justified, and who bears responsibility when predictions prove wrong. The retail pricing algorithm that contradicted industry wisdom, the pharmaceutical discovery system that identified promising compounds without explaining why, and the fraud detection models that flag threats human analysts miss all demonstrate a common reality that competitive advantage increasingly depends on machine learning systems that detect patterns beyond human cognitive capacity while operating beyond complete human comprehension.

This knowledge migration creates governance challenges that cannot be resolved through technical solutions alone. Organizations now possess systems that reliably predict customer behavior, market dynamics, and operational risks with statistical confidence that exceeds human intuition, yet those same systems cannot explain causal mechanisms, account for contextual factors outside their training data, or align their recommendations with organizational values they were never designed to recognize. The confidence paradox (where algorithmic certainty can suppress human judgment precisely when contextual understanding matters most) demands deliberate architectural choices about when machines should inform decisions versus when human authority must prevail.

The frameworks examined throughout this chapter provide starting points rather than complete answers. Risk-based classification systems like those in the EU AI Act establish legal boundaries around high-stakes decisions that cannot be fully automated regardless of algorithmic performance. Continuous monitoring protocols aligned with NIST guidance create feedback loops that detect when models drift, degrade, or encounter scenarios beyond their training scope. Human-in-the-loop, human-on-the-loop, and human-out-of-the-loop architectures offer structural options for calibrating oversight intensity to decision stakes and reversibility. Yet implementing these frameworks requires organizations to answer difficult questions that algorithms cannot resolve: Which business decisions involve values and relationships that transcend statistical optimization? When should executives override recommendations from systems with superior track records? How do companies preserve human expertise in domains where algorithms consistently outperform experienced professionals?

The organizational implications extend beyond individual decisions to institutional capabilities and culture. When critical business intelligence migrates from human minds to computational systems, companies face knowledge concentration risks, algorithmic de-skilling of employees who lose opportunities for experiential learning, and succession challenges that cannot be addressed through traditional mentorship and documentation. Addressing these vulnerabilities requires treating algorithmic knowledge as an organizational asset demanding active management, maintaining human centers of excellence even in automated domains, documenting not just technical specifications but business logic and constraints, and designing hybrid decision architectures where humans and machines contribute complementary capabilities.

The competitive pressures driving AI adoption in business intelligence show no signs of diminishing. Organizations that refuse to leverage pattern recognition algorithms will find themselves outmaneuvered by competitors who identify opportunities and optimize operations at scales impossible for human analysis alone. Yet the path forward requires more than technical implementation—it demands governance frameworks that preserve human accountability while capturing algorithmic advantages, organizational structures that maintain expertise even as automation expands, and leadership capable of exercising contextual judgment that statistical confidence cannot replace.

The question is not whether algorithms will continue detecting patterns humans cannot see, but whether organizations will build governance systems, accountability structures, and cultural norms necessary to ensure that superior pattern detection serves rather than subverts human values and strategic purpose. The next chapter examines how these governance challenges manifest when organizations move from leveraging AI insights to embedding algorithmic systems into formal decision-making authority, where questions of accountability, transparency, and ethical frameworks become not just operational concerns but fundamental requirements for maintaining stakeholder trust.

CHAPTER 3.

AI ETHICS AND GOVERNANCE: BUILDING FRAMEWORKS FOR ACCOUNTABILITY AND TRUST

A healthcare AI system recommended treatment protocols for over 200,000 patients before anyone discovered it had been systematically underestimating risk factors for a specific demographic group, resulting in suboptimal care recommendations that went unquestioned because the algorithm's accuracy metrics appeared excellent. The hospital's ethics committee, data science team, and clinical leadership had all approved the system, yet when patients experienced adverse outcomes, no individual or department could be held accountable exposing a fundamental flaw in how organizations approach AI governance.

This accountability vacuum represents one of the most pressing challenges facing organizations deploying AI decision-making systems today. As algorithms move from supporting human judgment to actively shaping critical choices, the traditional mechanisms for ensuring responsibility—clear reporting structures, documented decision processes, and individual accountability begin to break down. The question is no longer whether AI systems should be subject to ethical oversight, but how to construct governance frameworks that translate abstract principles into operational reality.

The challenge extends beyond identifying who should be held responsible when AI systems cause harm. Organizations must grapple with more fundamental questions about transparency, fairness, and trust in contexts where algorithmic decision-making often operates as a black box even to those who deploy it. A loan approval system may process thousands of applications with impressive efficiency, but when applicants from certain neighbors face systematically higher rejection rates, determining whether the algorithm reflects historical patterns, perpetuates existing biases, or identifies legitimate risk factors becomes extraordinarily complex. The technical team understands model architecture but may lack insight into social equity implications. The ethics committee comprehends fairness principles but cannot audit the algorithm's internal logic. Executive leadership bears ultimate responsibility but lacks the expertise to evaluate whether the system operates ethically.

This chapter examines how organizations and policymakers can move beyond aspirational ethics statements to build governance structures that ensure AI systems remain accountable to human values and societal expectations. Effective AI governance requires more than periodic audits or ethics review boards—it demands fundamental changes in organizational architecture, decision-making processes, and corporate culture. Companies must establish clear lines of responsibility that connect algorithmic outputs to human accountability, create transparency mechanisms that balance explainability with competitive concerns, and develop monitoring protocols that detect problems before they scale to affect thousands or millions of people.

The frameworks explored here draw from regulatory approaches across industries and jurisdictions, corporate governance models that have successfully integrated ethical oversight into AI deployment, and emerging best practices from organizations navigating these challenges in real time. These are not theoretical constructs but practical strategies being tested and refined by companies, government agencies, and institutions confronting the operational realities of AI accountability.

Some approaches emphasize regulatory compliance and external oversight. Others prioritize self-governance and industry standards. The most effective frameworks typically combine elements of both, recognizing that responsible AI governance requires collaboration between technologists who understand system capabilities, ethicists who can identify value conflicts, business leaders who must balance innovation with risk, and policymakers who establish societal guardrails.

The stakes extend beyond avoiding regulatory penalties or reputational damage. As AI systems increasingly influence decisions affecting employment, healthcare access, financial opportunity, and public safety, the governance frameworks we establish today will shape whether these technologies amplify human flourishing or exacerbate existing inequalities. Building accountability into AI systems from the beginning—rather than retrofitting ethics after deployment—represents the difference between technology that serves human values and technology that operates beyond meaningful human control. The path forward requires transforming ethics from philosophical debate into organizational practice, embedding accountability mechanisms into the technical and institutional architecture of AI decision-making itself.

The Accountability Architecture: Designing Organizational Structures That Assign Clear Responsibility for AI Decisions

Establishing clear accountability for AI decisions requires fundamentally rethinking organizational structures that evolved for human-driven processes. Traditional hierarchies assign responsibility through direct reporting relationships and documented decision chains A manager approves a loan, a physician orders a treatment, and an executive authorizes a strategic initiative. When AI systems enter these workflows, the clarity dissolves. The algorithm processes thousands of loan applications using criteria no single person fully understands. The clinical decision support tool recommends treatments based on pattern recognition across millions of patient records. The strategic planning system identifies market opportunities through analysis beyond human cognitive capacity. Suddenly, the question of who decided? Lacks a straightforward answer.

Organizations addressing this challenge effectively are building what governance researchers call accountability architectures —deliberate structural designs that map algorithmic outputs to human responsibility across the entire AI lifecycle.[38 13] These architectures recognize that AI decision-making involves multiple actors at different stages, each making choices that shape ultimate outcomes.[13] Data scientists select training datasets and model architectures. Product managers define use cases and performance metrics. Compliance officers establish guardrails and approval thresholds. Frontline employees interpret algorithmic recommendations and decide whether to apply or override them. Executive leadership authorizes deployment and allocates resources for monitoring and remediation.

The accountability architecture assigns specific duties to each role, creating what organizational theorists describe as responsibility mapping. This approach documents not just who builds or operates AI systems, but who bears accountability for fairness outcomes, who has authority to halt problematic deployments, who responds when individuals contest algorithmic decisions, and who ensures ongoing monitoring detects emerging harm before they scale. Without such mapping, organizations create what researchers call accountability systems where responsibility becomes so diffused across roles and processes that no individual can be held accountable when failures occur.[39]

Research on AI governance emphasizes that accountability requires more than assigning roles on paper. Those designated as responsible must possess genuine decision rights, which is the authority and practical capacity to understand, question, and alter AI-driven outcomes.[13] As organizational analysts examining large-scale automated systems have observed, the extent to which someone can change a decision determines precisely the extent to which they can be accountable for it.[39] This principle has profound implications for AI implementation. If a loan officer is held responsible for approvals but cannot access meaningful explanations of why the algorithm flagged certain applications, accountability becomes ceremonial rather than substantive. If a healthcare administrator bears responsibility for treatment recommendations but lacks authority to override the clinical AI or modify its parameters, the accountability assignment creates liability without control.

Effective accountability architectures therefore embed several structural features. They establish AI system owners who are specific individuals accountable for the end-to-end ethical and operational performance of particular applications. They designate fairness stewards responsible for monitoring distributional impacts across demographic groups and triggering reviews when disparities emerge.[13] They create cross-functional governance councils that bring together technical, legal, ethical, and domain expertise to review high-risk deployments before launch and investigate incidents after they occur. They implement escalation pathways that allow frontline staff and affected individuals to contest decisions and receive human review.[13]

Perhaps most critically, these architectures recognize that accountability cannot be retrofitted after deployment.[38] The organizational structures, role definitions, decision rights, and oversight mechanisms must be designed into AI systems from inception—integrated into requirements, embedded in workflows, and supported by training, incentives, and institutional culture that makes exercising accountability both possible and expected. The alternative is the accountability vacuum that leaves organizations legally exposed, stakeholders without recourse, and society unable to ensure AI systems serve rather than subvert human values.

From Principles to Practice: Translating Ethical Guidelines into Operational Governance Mechanisms

The gap between adopting ethical principles and implementing them operationally represents one of the most significant challenges in AI governance today. Organizations across sectors have embraced frameworks emphasizing fairness, transparency, accountability, and human oversight, yet translating these aspirations into concrete mechanisms that shape daily decisions remains elusive. A meta-analysis of AI ethics guidelines published since 2016 reveals hundreds of documents converging on similar values, yet most stop at abstract statements without specifying how to embed these principles into technical development, organizational processes, or oversight structures.[40]

This translation challenge stems partly from the inherent vagueness of ethical principles themselves. "Fairness" means different things across contexts—equal treatment, equal outcomes, proportional representation, or context-specific equity. "Transparency" might require explaining model logic, disclosing training data sources, or simply informing users that AI influenced a decision. Without operational definitions tied to specific contexts and measurable criteria, principles remain aspirational rather than actionable. Organizations must move from asking Should our AI be fair? To specifying what fair is according to which metric, measured how, with what acceptable variance, monitored by whom, and with what remediation when thresholds are breached.

The most effective governance mechanisms embed ethical requirements directly into the AI system lifecycle rather than treating ethics as a separate review layer.[40] This integration begins at problem formulation, where teams evaluate whether the proposed use case aligns with organizational values and human rights considerations before any technical work begins. During data collection, governance protocols enforce privacy standards, consent requirements, and representativeness criteria through mandatory documentation and approval gates. Model development incorporates fairness testing, robustness evaluation, and interpretability requirements as standard engineering practices rather than optional enhancements. Deployment triggers human oversight protocols, user communication requirements, and access controls. Post-deployment monitoring tracks performance drift, bias emergence, and stakeholder feedback through continuous measurement against predefined thresholds.

Standardized documentation tools have emerged as critical translation mechanisms. Model card structured templates describing an AI system's intended use, performance across demographic subgroups, limitations, and ethical considerations—convert transparency principles into concrete artifacts that enable informed decision-making. Similarly, datasheets for datasets document provenance, collection methods, known biases, and recommended uses, making abstract data governance principles tangible and auditable. These instruments create accountability by establishing clear records of choices made throughout development and deployment.

Algorithmic impact assessments represent another operational translation of ethical principles, particularly for high-risk applications.[40] These structured evaluations identify affected stakeholders, assess risks to fundamental rights and social equity, propose mitigation measures, and document consultation with domain experts or impacted communities. In jurisdictions with strong data protection frameworks, such assessments become legally mandated rather than voluntary, creating enforceable hooks that tie ethical principles directly to compliance obligations. The European Union's AI Act exemplifies this approach, transforming broad ethical concerns into risk-based regulatory requirements with specific technical and organizational obligations for high-risk systems, including mandatory risk management processes, data quality standards, logging capabilities, and human oversight measures.

Internal governance structures provide the institutional architecture for operationalizing principles.[40] Organizations establish AI ethics boards with clear mandates to review high-risk projects, cross-functional representation, and genuine decision authority. These are not merely advisory roles. Responsible AI offices develop policies derived from ethical principles, provide implementation guidance and tooling, coordinate risk assessments, and track incidents. These central functions work alongside distributed ownership models that assign explicit ethics and compliance responsibilities to product teams, data scientists, and engineers within their regular roles and performance evaluations.

The challenge extends beyond technical and organizational mechanisms to encompass power structures and institutional incentives. Scholars examining AI's political economy emphasize that governance frameworks risk legitimizing harmful practices if they address only model-level risks while ignoring extractive business models, surveillance practices, or systematic exclusion of marginalized communities from decision-making.[41] Effective translation of principles into practice therefore requires confronting not just how AI systems operate, but who designs them, who benefits from them, and whose values they ultimately serve.[41]

Transparency Without Compromise: Balancing Explainability Requirements with Competitive and Privacy Concerns

The demand for transparency in AI decision-making collides with two powerful countervailing forces, the competitive imperative to protect proprietary algorithms and the legal obligation to safeguard individual privacy. Organizations deploying AI systems face a genuine dilemma because stakeholders increasingly expect explanations of how algorithms reach decisions, yet providing detailed technical disclosures can expose trade secrets, enable system gaming, and inadvertently reveal sensitive personal information. Resolving this tension requires moving beyond the false choice between complete opacity and total transparency toward what governance researchers call purposeful transparency —providing the right information to the right actors for the right purposes.

The competitive concerns are not merely corporate excuses for avoiding scrutiny. Detailed model architecture, feature engineering techniques, and training methodologies often represent substantial investments and core competitive advantages. A financial institution's credit risk model, refined through years of development and millions in research expenditure, becomes vulnerable to replication if fully disclosed. Similarly, revealing algorithmic decision thresholds can enable strategic manipulation like applicants gaming loan approval systems, content creators exploiting platform recommendation algorithms, or adversaries crafting inputs designed to fool fraud detection models. Security researchers have demonstrated that excessive transparency about model internals can increase attack surfaces, making systems more vulnerable to adversarial manipulation.

Privacy constraints add another layer of complexity.[42] Explaining why an algorithm made a specific decision sometimes requires referencing individual-level data or revealing patterns that could expose sensitive attributes. A hiring algorithm's explanation might inadvertently disclose that certain demographic characteristics influenced outcomes, creating both privacy violations and discrimination evidence. Healthcare AI systems face particularly acute tensions as patients have legitimate interests in understanding treatment recommendations, yet detailed explanations could reveal sensitive medical information about other patients whose data contributed to the model's training or expose proprietary clinical knowledge.

The European Union's approach to algorithmic transparency illustrates the regulatory balancing act. The General Data Protection Regulation requires that individuals receive "meaningful information about the logic involved" when automated decisions produce legal or similarly significant effects, yet this obligation coexists with explicit protections for trade secrets and intellectual property.[42] The emerging EU AI Act takes this further, establishing risk-based transparency requirements where high-risk systems face stringent explainability obligations, while lower-risk applications receive more latitude to protect competitive information.[42] This tiered framework recognizes that transparency is not a binary choice but a spectrum calibrated to stakes and context.

Practical governance mechanisms have emerged to navigate these competing demands. Differentiated transparency provides varying levels of detail to different stakeholders—end users receive comprehensible explanations of factors affecting their individual decisions without technical minutiae, regulators and auditors gain access to detailed model documentation under confidentiality agreements, and internal governance teams maintain complete technical transparency for oversight purposes.[42 44] Model cards and algorithmic impact assessments offer standardized documentation that communicates system behavior, performance across demographic groups, and known limitations without revealing proprietary implementation details.[42]

Independent auditing under confidentiality represents another compromise instrument. Third-party auditors inspect full models and data practices, then issue public assurance statements about fairness, reliability, and compliance without disclosing competitive secrets. This approach, increasingly adopted in financial services and healthcare, enables external accountability while maintaining commercial viability. Organizations can demonstrate responsible governance through verifiable processes (bias testing protocols, monitoring systems, challenge mechanisms) rather than complete technical disclosure.[43]

The path forward requires recognizing that transparency serves accountability, not as an end itself but as a means to ensure AI systems remain subject to meaningful human oversight and correction.[44] The question is not whether organizations should reveal everything, but whether they provide sufficient information for affected individuals to understand and contest decisions, for regulators to verify compliance, and for society to hold deployers accountable when systems cause harm.[42 44]

The Audit Imperative: Establishing Continuous Monitoring and Assessment Protocols for AI Systems

Establishing accountability and transparency mechanisms represents only part of the governance challenge. Organizations must also implement continuous monitoring systems that detect problems as they emerge rather than discovering failures only after significant harm has occurred. The audit imperative recognizes that AI systems operate in dynamic environments where performance degrades, contexts shift, and unintended consequences manifest over time. A model that performs equitably at deployment may develop biases as data distributions change. An algorithm optimized for efficiency may gradually prioritize speed over accuracy as usage patterns evolve. Without systematic monitoring protocols, these degradations remain invisible until they scale to affect thousands of decisions.

The shift from one-time evaluation to continuous assessment reflects a fundamental recognition about AI systems in that they are not static products but evolving socio-technical assemblies whose behavior changes as data, contexts, and usage patterns shift.[40] Research on algorithmic accountability emphasizes that pre-deployment testing, while necessary, cannot anticipate all failure modes that emerge when systems encounter real-world complexity. A hiring algorithm tested on historical data may perform differently when labor market conditions change. A fraud detection system calibrated for normal transaction patterns may falter during economic disruptions. Continuous monitoring creates feedback loops that enable organizations to identify and address these emergent issues before they compound.

Effective monitoring protocols track multiple dimensions simultaneously. Performance monitoring measures whether systems maintain accuracy, precision, and reliability over time, detecting drift that signals degradation. Fairness monitoring examines outcomes across demographic groups and contextual categories, identifying disparities that may indicate bias even when overall performance metrics appear stable. Robustness monitoring assesses system behavior under edge cases, adversarial conditions, and distribution shifts that stress-test resilience. Security monitoring tracks unauthorized access attempts, data integrity issues, and manipulation efforts that could compromise system integrity.

These technical monitoring dimensions must integrate with organizational oversight mechanisms. Logging systems capture decision trails that enable retrospective analysis when contested outcomes or incidents occur.[40] Human review protocols establish thresholds that trigger manual examination like when confidence scores fall below specified levels, when decisions affect particularly sensitive contexts, or when patterns suggest systematic anomalies.[40] Escalation pathways connect frontline operators who observe concerning patterns to governance bodies with authority to investigate, modify, or suspend systems. Without such integration, technical monitoring generates data that never translates into corrective action.

The European Union's AI Act exemplifies regulatory frameworks that mandate continuous monitoring for high-risk applications. The legislation requires providers of high-risk AI systems to establish post-market monitoring plans that systematically collect and analyze data on system performance throughout the operational lifecycle. These obligations recognize that accountability cannot rest solely on pre-deployment assessments. It requires ongoing vigilance and adaptive response as systems operate in evolving environments. Similar expectations appear in sector-specific regulations governing healthcare algorithms, financial services models, and employment decision systems.

Organizations implementing robust monitoring face practical challenges around defining appropriate metrics, establishing meaningful thresholds, and allocating resources for sustained oversight. What constitutes acceptable performance variance before intervention becomes necessary? How should fairness metrics be calibrated for different contexts and stakeholder expectations?[41] [40] Who bears responsibility for reviewing monitoring outputs and authorizing remedial actions? These questions lack universal answers. Effective protocols require context-specific judgments that balance risk tolerance, operational constraints, and ethical commitments.

The audit imperative ultimately transforms AI governance from a compliance exercise into an operational discipline. Continuous monitoring embeds accountability into daily workflows rather than relegating ethics to periodic reviews. It creates institutional mechanisms for detecting when systems drift from intended behavior, when contexts change in ways that invalidate original assumptions, and when unintended consequences emerge that demand human intervention. Organizations that embrace this imperative build systems designed not for perfection but for ongoing scrutiny, correction, and alignment with human values across changing circumstances.[43]

Stakeholder Engagement Models: Building Trust Through Inclusive Governance and Meaningful Participation

Building accountability mechanisms and monitoring protocols addresses only part of the governance equation. Organizations must also determine who participates in shaping AI systems and how their input influences design, deployment, and oversight decisions. The question of stakeholder engagement cuts to the heart of democratic legitimacy in AI governance: when algorithms increasingly mediate access to credit, healthcare, employment, and public services, who gets to decide what values these systems embody, what trade-offs are acceptable, and what safeguards are necessary?

Traditional technology development has often treated stakeholder consultation as an optional enhancement or public relations exercise—gather feedback after core decisions are made, incorporate suggestions that don't require fundamental changes, and proceed with predetermined plans. This approach proves inadequate for AI systems that make consequential decisions affecting fundamental rights and life chances.[45] Research on AI governance emphasizes that meaningful stakeholder engagement must occur throughout the system lifecycle, from problem formulation through deployment and ongoing operation, with genuine influence over outcomes rather than ceremonial participation.[40 45]

The challenge begins with defining who counts as a stakeholder. The obvious answer includes individuals directly subject to algorithmic decisions—loan applicants, patients receiving treatment recommendations, job candidates evaluated by hiring systems, citizens affected by predictive policing tools. Yet the stakeholder landscape extends much further. Families and communities experience indirect impacts when AI systems shape resource allocation, employment opportunities, or public safety interventions in their neighborhoods. Workers whose roles are reorganized around algorithmic management have legitimate interests in how systems are designed and what human oversight remains. Civil society organizations representing marginalized groups bring crucial perspectives on how AI systems may perpetuate or amplify existing inequalities. Domain experts such as clinicians, social workers, and educators understand contextual factors that pure data analysis may miss. Regulators and policymakers must balance innovation incentives against public protection mandates.[40 45]

Effective engagement models recognize these diverse stakeholders require different participation mechanisms calibrated to their relationship with the system and the nature of decisions being made. Multi-stakeholder advisory bodies bring together representatives from government, industry, academia, and civil society to provide guidance on ethical frameworks, risk classifications, and policy options.[40 45] These bodies enhance legitimacy through diverse perspectives but risk becoming purely symbolic if recommendations lack enforcement mechanisms or if powerful interests dominate proceedings. Co-design and participatory design approaches involve affected communities directly in system development, using workshops, scenario building, and value-sensitive design methods to surface concerns and preferences that developers might otherwise overlook.[40 45] These intensive processes produce systems better aligned with community values but require substantial time and resources, along with careful facilitation to prevent more powerful voices from dominating.

Public consultation mechanisms —including comment periods, town halls, and citizens' assemblies—enable broader democratic input on controversial AI applications like facial recognition or automated welfare eligibility determination.[40 45] When well-designed with balanced expert input and adequate information, these processes bolster democratic legitimacy for high-stakes policy choices. However, their advisory nature means governments and organizations may selectively implement or ignore recommendations. Institutionalized ethics and oversight boards within organizations provide ongoing review of AI projects, assessing alignment with ethical guidelines and potentially blocking problematic deployments.[45] Their effectiveness depends critically on genuine decision authority, adequate resources, and independence from commercial pressures that might compromise ethical scrutiny.

The design of engagement processes matters as much as their existence. Meaningful participation requires procedural fairness, which includes accessible information in understandable language, engagement early enough to influence fundamental choices, and transparency about how input shapes decisions.[40 45] Power-sensitive design acknowledges that stakeholder processes unfold in contexts of asymmetric power, requiring deliberate representation of marginalized groups, independent facilitation, and capacity-building support that enables non-experts to participate effectively. Without such attention to process design, stakeholder engagement risks becoming what critics call "ethics washing"symbolic gestures that legitimize predetermined decisions without substantive change to practices, power structures, or outcomes.[40 45]

Regulatory Landscapes and Self-Governance: Navigating Compliance While Fostering Responsible Innovation

The tension between regulatory compliance and innovation acceleration represents one of the most consequential challenges facing organizations deploying AI decision-making systems. Executives often perceive regulation as constraint—bureaucratic overhead that slows development cycles, increases costs, and disadvantages them against competitors operating in less regulated jurisdictions. Yet this framing misses a fundamental reality that effective governance frameworks can enable rather than obstruct responsible innovation by establishing clear boundaries, reducing uncertainty, and building the stakeholder trust necessary for sustainable AI adoption.[46]

The regulatory landscape for AI governance varies dramatically across jurisdictions, creating complexity for organizations operating globally.[46 40] The European Union has moved toward comprehensive risk-based regulation through frameworks that impose stringent obligations on high-risk AI applications affecting fundamental rights, employment, and safety.[46 41 40] These approaches mandate conformity assessments, technical documentation, human oversight mechanisms, and post-market monitoring for systems deemed high-risk.[46 40] In contrast, the United States has favored sector-specific regulation and guidance documents that leverage existing legal frameworks such as anti-discrimination laws, consumer protection statutes, and financial services regulations rather than creating comprehensive AI-specific legislation.[40 47] Meanwhile, many jurisdictions are still developing their regulatory approaches, creating a patchwork of requirements that multinational organizations must navigate simultaneously.[46 40 47]

This regulatory fragmentation creates genuine operational challenges. A healthcare AI system approved under one jurisdiction's standards may face entirely different requirements elsewhere. A hiring algorithm that satisfies anti-discrimination testing in one market may be subject to different fairness metrics and transparency obligations in another. Organizations must either develop multiple compliance strategies for different markets or design systems that meet the most stringent requirements across all jurisdictions—an approach that increases development costs but simplifies governance architecture.

Yet regulatory compliance represents only the baseline for responsible AI governance. Organizations that treat ethics as synonymous with legal compliance risk missing broader accountability obligations to stakeholders whose trust determines long-term viability.[46] [43] Self-governance mechanisms—internal policies, ethics review boards, voluntary standards adoption, and proactive transparency practices—fill the gap between what regulation requires and what responsible deployment demands.[46] [41] These voluntary frameworks enable organizations to move faster than regulation evolves, addressing emerging ethical concerns before they become legal requirements or reputational crises.

The most sophisticated organizations are integrating compliance and self-governance into a unified framework that treats ethical considerations as design requirements rather than external constraints.[46] [41] This approach embeds legal and ethical review into product development workflows from inception, requiring teams to document intended use cases, assess potential harms, evaluate fairness across demographic groups, and establish monitoring protocols before deployment authorization.[46] [47] Rather than treating compliance as a gate that slows innovation, these organizations position governance as infrastructure that enables confident scaling—providing clear criteria for when systems are ready for broader deployment and establishing accountability mechanisms that protect both the organization and affected stakeholders.[46]

Industry standards and certification schemes are emerging as bridges between regulatory requirements and self-governance practices. Technical standards bodies are developing frameworks for AI system documentation, testing protocols, and risk management processes that organizations can adopt voluntarily yet which increasingly inform regulatory expectations.[46 47] These standards create common language across sectors and jurisdictions, reducing fragmentation while allowing flexibility for context-specific implementation.[40 47] Organizations that participate in standards development gain early insight into emerging regulatory directions while shaping frameworks that balance protection with operational feasibility.

The path forward requires reframing the relationship between regulation and innovation. Governance frameworks that establish clear accountability, provide mechanisms for redress, and build stakeholder trust do not obstruct innovation. Instead, they create the conditions under which AI systems can be deployed at scale without triggering backlash, moratoria, or catastrophic failures that set entire industries backward.[46 41 47] Organizations that embrace this perspective position governance not as compliance burden but as competitive advantage in markets where trust increasingly determines adoption.[43]The healthcare AI system that opened this chapter (operating undetected with systematic biases affecting 200,000 patients) illustrates why governance frameworks cannot remain aspirational. The accountability vacuum that emerged when adverse outcomes surfaced revealed a fundamental truth: organizations deploying AI decision-making systems require more than ethical principles and technical safeguards. They need operational architecture that assign clear responsibility, embed oversight into daily workflows, and create mechanisms for detecting and correcting problems before they scale to catastrophic proportions.

The frameworks examined throughout this chapter represent practical responses to this imperative, drawn from organizations navigating AI governance challenges in real time. Accountability architectures map algorithmic outputs to human responsibility across the entire system lifecycle. Operational mechanisms translate abstract fairness principles into concrete testing protocols, documentation requirements, and review gates. Transparency approaches balance explainability demands with legitimate competitive and privacy concerns. Continuous monitoring systems detect performance degradation, bias emergence, and unintended consequences as they develop rather than after significant harm occurs. Stakeholder engagement models incorporate diverse perspectives into system design and oversight rather than treating consultation as ceremonial validation.

These governance elements share a common recognition that AI systems operate as socio-technical assemblies whose behavior emerges from the interaction of algorithms, data, organizational processes, and human decisions. Effective governance must therefore address not just model architecture or training data quality, but the institutional structures, incentive systems, and cultural norms that shape how AI systems are developed, deployed, and monitored. Technical audits matter, but so do organizational readiness assessments. Algorithmic fairness testing is essential, but insufficient without clear escalation pathways that connect frontline observations to governance bodies with authority to intervene. Pre-deployment impact assessments provide crucial safeguards, yet they cannot anticipate all failure modes that emerge when systems encounter real-world complexity.

The regulatory landscape adds another dimension to governance challenges, creating both constraints and opportunities for organizations navigating AI adoption. Jurisdictions worldwide are developing divergent approaches—comprehensive risk-based frameworks in some regions, sector-specific regulations in others, and emerging standards in many. This fragmentation creates operational complexity for multinational organizations yet also drives convergence around core principles—transparency appropriate to risk levels, human oversight for consequential decisions, mechanisms for redress when systems cause harm, and ongoing monitoring rather than one-time approval. Organizations that treat these regulatory requirements as minimum baselines while building more comprehensive self-governance frameworks position themselves to move confidently as expectations evolve.

Perhaps most critically, the governance frameworks explored here challenge the false dichotomy between innovation and accountability. The perception that ethical oversight necessarily slows development or constrains competitive advantage misses a fundamental reality about sustainable AI adoption. Systems deployed without adequate governance create the conditions for catastrophic failures that trigger regulatory backlash, erode stakeholder trust, and ultimately constrain entire industries. The insurance company that opened this chapter discovered this truth through millions in settlements and irreparable reputational damage. Organizations that embed accountability from inception, treating governance as infrastructure rather than obstacle build the trust necessary for scaling AI systems across sensitive domains.

The path forward requires transforming AI ethics from philosophical debate into operational discipline. This transformation demands collaboration across the stakeholder landscape examined throughout this chapter via technologists who understand system capabilities and limitations, ethicists who identify value conflicts and fairness concerns, business leaders who balance innovation imperatives with risk management, policymakers who establish societal guardrails, and affected communities whose lived experiences reveal consequences that data analysis alone cannot capture. The governance frameworks that emerge from this collaboration will determine whether AI decision-making systems amplify human flourishing or operate beyond meaningful human control—making the architecture of accountability not a technical detail but a foundational question about the future we choose to build.

CHAPTER 4.

STRATEGIC AI IMPLEMENTATION: BALANCING INNOVATION WITH HUMAN AGENCY

A financial services firm's AI-powered investment system generated returns 12% above industry benchmarks for eighteen months, operating with minimal human intervention as executives celebrated their technological advantage. When market conditions shifted unexpectedly, the system continued executing trades based on patterns that no longer applied, and by the time human analysts recognized the problem and understood how to intervene, the firm had lost $340 million—not because the AI failed technically, but because the implementation design had optimized for speed over human agency. This cautionary example reveals a fundamental issue facing organizations today: how to capture AI's competitive advantages without creating systems that operate beyond meaningful human control.

The promise of AI-driven decision-making is undeniably compelling. Organizations implementing these systems report significant efficiency gains, improved accuracy in pattern recognition, and the ability to process information at scales impossible for human teams. Yet the path from pilot program to enterprise-wide deployment is littered with expensive failures, ethical breaches, and systems that delivered technical performance while undermining the very human judgment they were meant to enhance. The difference between successful and disastrous AI implementation rarely comes down to algorithmic sophistication but rather hinges on strategic choices about where humans remain in the decision-making loop, how override mechanisms function, and which decisions should never be fully automated regardless of technical capability.

This chapter examines the practical architecture of responsible AI implementation, moving beyond abstract principles to address the concrete design choices that determine whether AI systems augment or replace human agency. The challenge facing executives, consultants, and policymakers is not simply whether to adopt AI decision-making capabilities, but how to structure that adoption in ways that preserve meaningful human oversight while still capturing the speed and analytical advantages that make AI valuable. These are not purely technical questions. They require strategic thinking about organizational values, risk tolerance, and the fundamental nature of human-machine collaboration.

The competitive pressure to automate decision-making creates a powerful incentive to minimize human intervention, which algorithms can perceive as friction that slows processes and reduces efficiency. Organizations that succumb to this pressure often discover too late that the speed advantages came at the cost of adaptability, contextual judgment, and the ability to recognize when algorithmic recommendations no longer align with changing circumstances. Conversely, organizations that maintain excessive human oversight may fail to capture AI's benefits, creating expensive systems that employees circumvent because the implementation design made them cumbersome rather than enabling.

Finding the appropriate balance requires understanding that different decision types demand different levels of human involvement. Routine operational choices with clear parameters and immediate feedback loops may function effectively with minimal human oversight, while strategic decisions with long-term consequences, ethical implications, or high uncertainty require architectures that preserve robust human agency. The implementation challenge lies in designing systems that match oversight intensity to decision significance while creating clear intervention points where human judgment can effectively supersede algorithmic recommendations.

This chapter provides frameworks for making these critical implementation choices, examining how organizations can assess their readiness for AI adoption, design systems with appropriate human-in-the-loop mechanisms, and establish the cultural and structural prerequisites for responsible deployment. Through analysis of both successful implementations and instructive failures across industries, we explore how to transform AI from a source of competitive anxiety into a strategic capability that enhances rather than diminishes human decision-making authority. The goal is not to slow innovation but to ensure that the rush to automate doesn't inadvertently create systems that operate beyond our ability to understand, question, or control them when circumstances demand human intervention.

The Human-in-the-Loop Spectrum: Designing Appropriate Levels of Human Oversight for Different Decision Types

Not all decisions require the same level of human involvement, yet organizations often default to either full automation or excessive oversight without systematically evaluating what each decision type demands. The concept of a human-in-the-loop spectrum provides a structured approach to matching oversight intensity with decision characteristics, moving beyond binary choices to recognize that AI and humans can collaborate in fundamentally different ways depending on context, risk, and the nature of judgment required.[51]

At one end of this spectrum sits human-in-the-loop (HITL) decision-making, where humans retain direct authority to approve, modify, or reject AI recommendations before any action occurs.[48] In HITL architectures, AI systems analyze data and generate preliminary assessments, but a human decision-maker must actively engage with each recommendation, applying domain expertise, contextual knowledge, and ethical considerations before finalizing the choice.[48] Healthcare diagnostics exemplify this model. AI systems analyze medical imaging to flag potential malignancies, but radiologists integrate these suggestions with clinical data and patient history before rendering diagnoses.[48 51] The human remains the final decision-maker, with AI functioning as a powerful analytical assistant rather than an autonomous agent.[48]

HITL arrangements offer critical advantages for high-stakes decisions. They enable improved accuracy by combining AI's pattern recognition capabilities with human contextual understanding, particularly for outlier cases where algorithmic models may falter.[48 52] They also preserve transparency and accountability, since human evaluators can demand explanations and document their reasoning.[48] Research demonstrates that in domains like radiology, the combination of human expertise and AI assistance outperforms either humans or algorithms working alone, precisely because each compensates for the other's weaknesses.[51] However, HITL models sacrifice speed and scalability, making them impractical for high-volume, time-sensitive decisions where individual choices carry lower risk.

The middle ground is occupied by human-on-the-loop (HOTL) systems, where AI operates with substantial autonomy while humans maintain supervisory oversight and the ability to intervene or override.[49 50] Rather than approving each decision, human operators monitor system performance, adjust parameters, and step in when circumstances exceed the AI's design boundaries. Financial institutions employ this model for fraud detection. Machine learning systems automatically flag suspicious transactions and may block low-risk cases autonomously, while human analysts review borderline situations and maintain authority to adjust detection thresholds or halt automated responses when patterns shift unexpectedly.[48]

HOTL architectures enable machine-speed operation in dynamic environments while preserving human authority to redirect or halt systems when needed.[50] Military command and control systems increasingly adopt this approach for rapid-response scenarios, where autonomous modules generate tactical options but human subject-matter experts select among alternatives and supervise execution.[50] The Joint Air Power Competence Centre's analysis of these systems emphasizes "management by consent," requiring operators to periodically exercise active judgment rather than passively monitoring until alarms trigger—a design choice intended to counteract automation bias and maintain genuine human engagement.[50]

At the spectrum's far end, human-out-of-the-loop (HOOTL) systems sense, decide, and act without human involvement in individual decisions.[49] These fully autonomous systems operate at machine speed, handling high-volume tasks where individual errors are tolerable or can be mitigated through downstream controls.[49] Email spam filtering and basic recommendation engines exemplify HOOTL applications where the cost of occasional errors is low, and the volume makes human review of each decision impractical.

The challenge facing organizations is not selecting a single model but deliberately mapping different decision types to appropriate points on this spectrum. Decisions involving fundamental rights, irreversible consequences, or significant ethical dimensions demand HITL oversight regardless of efficiency costs.[48 52] Time-critical operational choices with moderate risk profiles may justify HOTL arrangements.[49 50] Routine, low-impact decisions can operate as HOOTL systems, provided robust monitoring and periodic audits detect systematic problems before they compound.[49 52] The EU AI Act operationalizes this risk-proportionality principle by requiring stronger human oversight for high-risk AI applications, effectively mandating that regulatory compliance follows the spectrum logic rather than treating all automation identically.[52]

Implementation Architecture: Building Systems That Preserve Meaningful Human Intervention Points

The technical architecture of AI systems determines whether human intervention remains genuinely possible or becomes merely theoretical. Organizations frequently implement AI decision-making platforms with nominal human oversight mechanisms that prove ineffective when circumstances demand actual intervention, creating what researchers call accountability gaps—situations where responsibility diffuses across technical systems, organizational hierarchies, and automated processes until no one can effectively exercise judgment or control outcomes.

Effective implementation architecture requires deliberate design choices that embed intervention points throughout the decision-making pipeline rather than treating human oversight as a final checkpoint. Research from the Partnership on AI emphasizes that meaningful human control depends on three architectural elements: transparency about how systems generate recommendations, accessibility of intervention mechanisms that humans can actually operate under realistic conditions, and reversibility that allows decisions to be reconsidered when context changes or errors emerge. Systems lacking any of these elements may technically include humans in the process while functionally operating beyond genuine human authority.

The placement of intervention points matters as much as their existence. AI systems that present recommendations as final outputs with binary approve-reject options create fundamentally different human roles than systems that expose intermediate reasoning steps and allow humans to adjust parameters, question assumptions, or request alternative analyses. A credit approval system that shows an applicant's risk score and asks for yes-no confirmation provides minimal intervention capability compared to one that displays which factors most influenced the assessment, allows underwriters to weight contextual information the algorithm cannot access, and enables case-by-case adjustments based on circumstances the training data never encountered.

The European Union's proposed AI Act recognizes this architectural imperative by requiring that high-risk AI systems be designed to enable human oversight through measures that include providing humans with a proper understanding of system capacities and limitations, enabling humans to correctly interpret system output, and allowing humans to decide not to use the system or to override its output. These requirements acknowledge that intervention points must be designed into system architecture from the beginning rather than added as afterthoughts, since retrofitting meaningful human control into systems optimized for autonomous operation rarely succeeds.

Implementation architecture must also address the temporal dimension of intervention. Real-time decisions demand intervention mechanisms that humans can operate within the available timeframe, while systems making irreversible choices require stronger safeguards than those whose outputs can be reviewed and corrected. A hiring algorithm that screens thousands of applications can reasonably include human review of candidates the system advances, but an autonomous trading system executing millisecond transactions requires different architectural approaches—perhaps limiting the system's operational boundaries, implementing automatic circuit breakers, or restricting its use to contexts where speed justifies reduced human control.

The challenge increases when competitive pressure incentivizes minimizing human intervention as friction that reduces efficiency.

Organizations implementing AI systems often discover that meaningful oversight slows processes, increases costs, and requires maintaining human expertise that automation was supposed to eliminate. This creates powerful incentives to design intervention points that satisfy compliance requirements while remaining too cumbersome for regular use—a phenomenon researchers' term "oversight theater," where human review exists formally but rarely influences actual outcomes.

Preserving genuine intervention capability requires organizational commitment beyond technical design. Systems must be architected so that humans can understand recommendations well enough to evaluate them critically[53], possess sufficient authority to override algorithmic outputs without bureaucratic obstacles[53], and face accountability structures that reward appropriate intervention rather than penalizing decisions that slow automated processes. Without these supporting elements, even well-designed technical intervention points become vestigial features that humans learn to ignore, transforming oversight from meaningful control into procedural formality that provides legal cover while abandoning substantive human agency.

The Override Challenge: Creating Effective Mechanisms for Human Judgment to Supersede Algorithmic Recommendations

The ability to override algorithmic recommendations represents one of the most critical yet poorly implemented elements of responsible AI deployment. Organizations frequently design systems with nominal human override capabilities that prove ineffective when circumstances demand actual intervention and subsequently create accountability gaps.[31 55] The challenge lies not simply in providing an override button, but in architecting systems where human judgment can genuinely supersede algorithmic outputs when context, values, or unforeseen circumstances demand it.[31 55]

Effective override mechanisms must address three fundamental design requirements: visibility, accessibility, and organizational support. Visibility means humans can understand what the algorithm recommends and why, with sufficient transparency to evaluate whether the recommendation aligns with broader objectives and ethical considerations.[54 32] Without interpretable explanations tailored to the decision-maker's domain expertise, override becomes an uninformed guess rather than reasoned judgment.[54] A credit approval system that displays only a risk score provides minimal override capability compared to one that shows which factors most influenced the assessment and allows underwriters to weight contextual information the algorithm cannot access.

Accessibility requires that override controls remain practically usable rather than merely formally available. Many systems bury override functions in complex interfaces, require lengthy justifications that discourage use, or create approval processes so cumbersome that time-pressured professionals' default to accepting algorithmic outputs. Research on human-machine decision-making demonstrates that when override mechanisms impose significant friction, humans tend toward automation bias and accept recommendations even when professional judgment suggests otherwise.[31] Effective architectures place override control prominently, require reasonable documentation that captures learning value without deterring responsible use, and ensure that exercising override authority remains feasible within realistic operational timeframes.

Organizational support determines whether override mechanisms function as genuine decision-making tools or become vestigial features that satisfy compliance requirements while rarely influencing outcomes. When productivity metrics implicitly penalize professionals who slow processes by questioning algorithmic recommendations, or when organizational culture treats override as insubordination rather than legitimate professional judgment, even well-designed technical mechanisms fail.[31 32] Nimrod Partush, a data science executive in cybersecurity, emphasizes that "humans should almost always possess the ability to overrule AI decisions," yet this capability requires institutional frameworks that recognize override as an expected element of professional responsibility rather than deviation from standard procedure.[55]

The appropriate strength and placement of override mechanisms varies systematically with decision characteristics. High-stakes choices with irreversible consequences such as medical diagnoses, employment terminations, and criminal sentencing demand robust pre-decision human authority regardless of efficiency costs.[31] Time-critical operational decisions with moderate risk profiles may justify post-hoc override arrangements where algorithms act first but humans can rapidly reverse or modify outcomes, as seen in fraud detection systems or autonomous emergency braking.[31] The key insight from decision-intelligence research is that there exists no universal override model, and effective architectures match oversight intensity to the specific combination of reversibility, time-sensitivity, and potential harm that characterizes each decision type.[31 32]

Organizations that treat overrides as obstacles to efficiency rather than essential safeguards discover too late that speed advantages came at the cost of adaptability and contextual judgment. Conversely, systems designed with genuine override capability create valuable feedback loops. When and why humans disagree with algorithms provides critical data for refining both models and governance frameworks.[31 32] The override challenge ultimately requires recognizing that human judgment superseding algorithmic recommendations is not a failure of automation but a fundamental feature of responsible AI implementation, transforming override from technical afterthought into deliberate architectural choice that preserves human agency while capturing AI's analytical advantages.

Organizational Readiness: Assessing Cultural and Structural Prerequisites for Responsible AI Adoption

Before organizations can successfully implement AI decision-making systems with appropriate human oversight, they must honestly assess whether their culture and structure can support responsible adoption. The most sophisticated algorithms and carefully designed intervention mechanisms fail when deployed into organizational environments unprepared for the fundamental shifts that AI integration demands. Research from decision intelligence frameworks demonstrates that successful AI adoption requires simultaneous readiness across multiple dimensions: strategic clarity about AI's role, leadership understanding of capabilities and limitations, technical infrastructure, data governance maturity, workforce skills, and cultural alignment with data-driven decision-making.[1]

Strategic readiness begins with leadership articulating a coherent vision for how AI will enhance decision-making while maintaining human accountability.[1] This extends beyond viewing AI as a technological upgrade to recognizing it as a transformational force that will reshape organizational processes, roles, and responsibilities. Organizations must define specific decision domains where AI will be deployed, establish clear metrics for success, and determine appropriate levels of human involvement for different decision types.[1] Without this strategic foundation, AI implementations become technology-driven experiments rather than purposeful enhancements to organizational capability.

The cultural dimension of readiness proves particularly challenging because it requires shifts in deeply embedded organizational behaviors and beliefs. Successful AI integration demands workforce acceptance of data-driven decision-making, which may conflict with existing cultures that value intuition-based judgment or hierarchical authority. Employees accustomed to making decisions based on experience and instinct must develop comfort with algorithmic recommendations while maintaining healthy skepticism about machine outputs. This cultural transformation requires psychological safety. Workers need assurance that AI augmentation does not threaten their employment or autonomy, and that questioning algorithmic recommendations represents professional responsibility rather than resistance to innovation.

Organizations must also cultivate ethical consciousness throughout their workforce. This includes awareness of how algorithmic decisions might discriminate against certain populations, capacity to identify and mitigate bias, and commitment to ensuring AI systems serve legitimate organizational and social purposes. The challenge intensifies in high-stakes domains where algorithmic errors carry profound consequences. Criminal justice systems, for example, have deployed AI algorithms with approximately 30% error rates to guide consequential decisions about bail and sentencing[56]—implementations that proceeded without adequate assessment of whether organizations possessed the ethical frameworks and oversight capabilities to use such systems responsibly.

Structural readiness encompasses the operational capabilities required to implement and maintain AI systems effectively. Organizations need robust data governance structures that ensure data quality, since poor data leads to biased algorithms and unreliable decision support.[1] They require technical infrastructure capable of supporting AI systems and integration with existing enterprise processes. Perhaps most critically, they need monitoring and evaluation systems that continuously assess

AI performance, detect model drift, identify emerging biases, and implement corrective actions. Without these structural elements, organizations cannot maintain the ongoing oversight that responsible AI adoption demands.

Talent availability represents another critical prerequisite. Successful AI implementation requires diverse skill sets: data scientists to build systems, domain experts who understand business context and can validate recommendations, change management professionals to guide organizational transitions, and ethics specialists to identify potential harms.[1] Few organizations currently possess all necessary capabilities internally, requiring strategic decisions about hiring, partnerships, or phased implementation that matches ambition to actual capacity.

The organizations that succeed in responsible AI adoption are those that conduct honest readiness assessments across all these dimensions before deployment, invest in building missing capabilities, and recognize that AI integration represents a fundamental transformation requiring change at all organizational levels.[57] Those that focus exclusively on technical implementation while neglecting cultural, ethical, or structural prerequisites discover that even sophisticated algorithms cannot overcome organizational unreadiness for the human-machine collaboration that effective AI decision-making requires.

The Competitive Pressure Paradox: Maintaining Human Agency When Speed and Efficiency Drive Market Advantage

The most insidious threat to human agency in AI implementation comes not from technical limitations but from market dynamics that reward speed over deliberation. Organizations face a structural dilemma when competitors deploying AI systems with minimal human oversight gain measurable advantages in response time, operational efficiency, and cost reduction, creating pressure to match or exceed their automation levels regardless of agency concerns. This competitive pressure paradox transforms responsible AI adoption from a strategic choice into what many executives perceive as a luxury they cannot afford, which is a perception that fundamentally misunderstands both the nature of sustainable competitive advantage and the long-term costs of agency erosion.

Research on AI adoption patterns across industries reveals that organizations frequently justify reduced human oversight by citing competitive necessity rather than technical appropriateness. When one financial institution automates loan approvals to deliver decisions in minutes rather than days, competitors face pressure to match that speed or lose market share to customers who prioritize convenience. When retailers deploy dynamic pricing algorithms that adjust thousands of prices hourly based on demand signals, rivals must implement similar systems or accept margin disadvantages. The logic appears compelling. Human review introduces latency that customers experience as friction and markets punish as inefficiency.

Yet this framing obscures a critical distinction between decisions where speed creates genuine value and those where velocity serves primarily to preempt human judgment. Automated fraud detection that blocks suspicious transactions in milliseconds protects both institutions and customers in ways human review cannot match. In this case, the speed directly serves the decision's purpose. Conversely, employment screening systems that reject applications in seconds rather than days serve organizational efficiency more than decision quality, since hiring outcomes depend on contextual judgments that algorithms struggle to capture and speed rarely improves. The competitive pressure paradox emerges when organizations treat these fundamentally different decision types identically and apply the same automation logic to both and measure success primarily through efficiency metrics.

The challenge grows because competitive dynamics create what game theorists recognize as a coordination problem.[59] Individual organizations that maintain robust human oversight may operate at disadvantages relative to competitors who automate more aggressively, yet when all players minimize human agency, the collective outcome can be worse for everyone—reduced trust, increased regulatory scrutiny, and systemic risks that no single actor can manage alone. Financial markets experienced this dynamic during the 2010 "Flash Crash," when algorithmic trading systems interacting at machine speed produced a market collapse that erased nearly $1 trillion in value within minutes, demonstrating how optimization for individual speed advantages can generate collective catastrophe.

Maintaining human agency under competitive pressure requires reframing what constitutes sustainable advantage. Organizations that preserve meaningful human oversight in high-stakes decisions may sacrifice short-term efficiency gains but develop capabilities that prove more durable such as institutional knowledge about when algorithms fail, workforce expertise that prevents skill atrophy, stakeholder trust that becomes valuable when competitors face backlash, and adaptive capacity to recognize when market conditions exceed algorithmic training parameters. The financial services firm that lost $340 million when its AI system continued executing outdated strategies possessed technical sophistication but lacked the human engagement architecture necessary to intervene effectively when circumstances shifted failure not of algorithms but of implementation design that prioritized speed over agency.

The paradox resolves not by rejecting competitive pressure as illegitimate but by recognizing that human agency itself represents a strategic capability rather than operational overhead.[58 59] Organizations that architect AI systems to preserve meaningful human intervention points, maintain workforce competencies for critical judgment, and cultivate cultures where questioning algorithmic outputs constitutes professional responsibility rather than insubordination develop competitive advantages that pure automation cannot replicate. And those advantages become increasingly valuable as AI adoption matures and markets begin distinguishing between systems that serve genuine value creation and those that merely optimize for speed.

Strategic Boundaries: Identifying Decisions That Should Never Be Fully Automated Regardless of Technical Capability

Even as organizations develop sophisticated frameworks for human-in-the-loop oversight and intervention mechanisms, a more fundamental question demands attention: which decisions should never be fully automated regardless of how technically capable AI systems become? This question shifts the focus from *how* to implement AI responsibly to *where* strategic boundaries must be drawn for decisions that, by their nature, require irreplaceable human judgment, moral responsibility, and accountability that machines cannot legitimately provide.

The distinction between technical capability and moral permissibility becomes critical as AI systems demonstrate increasing proficiency across decision domains.[32 34] Algorithms can now match or exceed human performance in medical diagnosis, legal research, credit assessment, and even some aspects of strategic planning. Yet performance metrics alone provide insufficient guidance for determining automation boundaries. A system that achieves 95% accuracy in predicting recidivism rates may be technically impressive while remaining ethically inappropriate for determining criminal sentences, because the decision involves contested values, fundamental rights, and the requirement for a morally accountable human who can justify choices and bear responsibility for errors.[34]

Research in decision intelligence and AI governance identifies several characteristics that mark decisions as inappropriate for full automation.[1] [31] [32] [34] Decisions involving fundamental human rights and dignity — determinations about bodily integrity, liberty, legal personhood, or life and death—demand human judgment because they engage non-quantifiable values and moral frameworks that algorithms cannot authentically navigate.[31] [34] When immigration systems use AI to assess asylum applications, the technical capability to process applications efficiently does not eliminate the moral requirement for human adjudicators who can weigh complex narratives, understand context that training data never captured, and be held accountable for decisions that shape entire life trajectories.

Irreversibility and catastrophic risk constitute another boundary criterion.[1] [32] Decisions whose errors produce irreversible harm or existential consequences like military escalation, critical infrastructure governance, or deployment of powerful AI systems whose societal impacts remain uncertain must retain robust human authority regardless of algorithmic sophistication. The 2010 Flash Crash demonstrated how algorithmic trading systems optimized for speed could interact to produce market collapse by erasing nearly one trillion dollars in minutes. The technical capability for machine-speed trading did not justify removing human authority to halt systems when collective behavior exceeded design parameters.

Decisions that are fundamentally value-laden rather than fact-determined present a third category requiring human primacy.[34] Public policy trade-offs between liberty and security, fairness and efficiency, or individual rights and collective welfare involve choosing between legitimate but conflicting values. AI can model scenarios and project consequences, but cannot legitimately resolve such conflicts because it possesses no intrinsic purposes, commitments, or moral standing.[34]

When healthcare systems must allocate scarce life-saving resources during crises, algorithms may inform triage protocols, but the ethical choices about whose lives to prioritize demand human deliberation, transparent reasoning, and accountable decision-makers who can be questioned and challenged.

The requirement for moral responsibility and answerability creates practical boundaries around automation.[34] Criminal sentencing, professional ethical judgments, and organizational decisions with profound human consequences require decision-makers who can articulate reasoning, express appropriate moral emotions, and bear social and legal consequences for errors. AI systems lack consciousness, understanding, and capacity for genuine moral responsibility. They can only redistribute accountability across designers, deployers, and overseers, often creating accountability gaps where no one can effectively answer for outcomes.[34]

These strategic boundaries do not reject AI's role in such decisions entirely. Rather, they establish that in domains touching fundamental rights, irreversible consequences, contested values, or requirements for moral answerability, AI must remain advisory while final authority resides with identifiable humans who can justify their judgments and accept responsibility for outcomes. The challenge for organizations and policymakers lies in translating these principles into operational taxonomies that classify decisions by risk and ethical stakes, then designing governance structures that enforce appropriate boundaries even when competitive pressure incentivizes further automation.[1] [32]The financial services firm that lost $340 million when market conditions shifted unexpectedly illustrates a fundamental truth about AI implementation: technical sophistication means nothing without strategic architecture that preserves human agency. Organizations rushing to capture AI's competitive advantages often discover too late that speed and efficiency gains came at the cost of adaptability, contextual judgment, and the ability to intervene when circumstances demand human control. The difference between successful and disastrous AI adoption rarely hinges on algorithmic performance—it depends on deliberate design choices about where humans remain meaningfully engaged, how override mechanisms actually function under pressure, and which decisions should never be fully automated regardless of technical capability.

The human-in-the-loop spectrum provides a structured framework for matching oversight intensity to decision characteristics, moving beyond binary automation choices to recognize that different contexts demand fundamentally different collaboration models. High-stakes decisions involving irreversible consequences or fundamental rights require robust human authority regardless of efficiency costs, while routine operational choices with clear parameters may function effectively with supervisory oversight or minimal intervention. The critical insight is that no universal model exists. Effective implementation requires systematically evaluating each decision type against criteria including reversibility, time-sensitivity, ethical implications, and potential harm, then architecting appropriate intervention points throughout the decision-making pipeline.

Yet intervention points prove meaningless unless humans can actually use them when circumstances demand. Override mechanisms fail when they lack visibility into algorithmic reasoning, impose cumbersome processes that discourage use, or exist within organizational cultures that treat questioning AI recommendations as obstruction rather than professional responsibility. Effective architecture ensures that humans understand what systems recommend and why, can access override controls within realistic operational timeframes, and operate in institutional environments that recognize intervention as expected professional judgment rather than deviation from standard procedure. Without these supporting elements, even well-designed technical mechanisms become vestigial features satisfying compliance requirements while abandoning substantive human agency.

The competitive pressure paradox represents perhaps the most insidious threat to responsible implementation. Market dynamics reward organizations that minimize human oversight to maximize speed and efficiency, creating structural incentives to automate aggressively regardless of agency concerns. This pressure transforms responsible AI adoption into what many executives perceive as a luxury they cannot afford—a perception that fundamentally misunderstands both sustainable competitive advantage and the long-term costs of agency erosion. Organizations that preserve meaningful human engagement develop capabilities that prove more durable than pure automation: institutional knowledge about when algorithms fail, workforce expertise that prevents skill atrophy, stakeholder trust that becomes valuable when competitors face backlash, and adaptive capacity to recognize when circumstances exceed algorithmic training parameters.

Strategic boundaries must be drawn around decisions that should never be fully automated regardless of technical capability. Choices involving fundamental human rights, irreversible consequences, contested values, or requirements for moral answerability demand human primacy because they engage non-quantifiable considerations that algorithms cannot authentically navigate and require accountable decision-makers who can justify choices and bear responsibility for errors. Technical performance metrics alone provide insufficient guidance. A system achieving impressive accuracy remains ethically inappropriate for certain domains because the decision's nature demands irreplaceable human judgment and moral responsibility that machines cannot legitimately provide.

The path forward requires recognizing that AI implementation represents not merely a technical project but a strategic transformation requiring simultaneous readiness across cultural, structural, and ethical dimensions. Organizations that conduct honest assessments of their preparedness, invest in building missing capabilities, and architect systems that preserve human agency while capturing AI's analytical advantages position themselves for sustainable success in an increasingly automated world. The goal is not to slow innovation but to ensure that the rush to automate creates systems that enhance rather than diminish human decision-making authority and thereby transforming AI from a source of competitive anxiety into a strategic capability that serves rather than supplants human judgment.

CHAPTER 5.

HUMAN-AI COLLABORATION: REDEFINING LEADERSHIP IN THE AGE OF INTELLIGENT AUTOMATION

The CEO of a multinational logistics company sat across from her executive team, reviewing an AI system's recommendation to restructure operations in a way that would eliminate 3,000 jobs while increasing efficiency by 23%. The algorithm's analysis was impeccable, the financial projections compelling, yet the data couldn't capture the company's commitment to communities where it had operated for decades, the knowledge held by experienced workers, and the long-term reputation costs. Those factors demanded a leadership judgment that no machine could render. This scenario, playing out in boardrooms worldwide, reveals a fundamental truth about the AI transformation: technology may augment decision-making capabilities, but it cannot replace the distinctly human dimensions of leadership that determine organizational direction and values.

The emergence of AI as a collaborative partner in strategic decision-making represents not just a technological shift but a redefinition of what leadership means in modern organizations. Executives who once relied primarily on human advisors, market research, and institutional knowledge now find themselves interpreting algorithmic recommendations, orchestrating hybrid human-machine teams, and making judgment calls about when data-driven insights should be overridden by considerations that algorithms cannot quantify. This evolution demands new competencies that few leadership development programs have addressed: the ability to understand AI capabilities and limitations without technical expertise, the discernment to recognize which decisions require human judgment regardless of algorithmic confidence, and the skill to cultivate organizational cultures that leverage machine intelligence while preserving human creativity, empathy, and ethical reasoning.

The challenge extends beyond individual leader development to encompass fundamental questions about organizational intelligence itself. When both humans and machines contribute insights to strategic decisions, traditional hierarchies and decision-making processes require reconsideration. Who holds authority when an algorithm's recommendation contradicts executive intuition? How do leaders maintain strategic vision when AI systems identify patterns and opportunities that humans cannot perceive? What happens to organizational learning and knowledge transfer when critical intelligence resides within algorithmic systems rather than experienced personnel? These questions demand answers not in abstract theory but in practical frameworks that leaders can implement immediately as AI capabilities expand.

This chapter examines how leadership must evolve to meet the demands of human-AI collaboration, moving beyond the false choice between embracing automation uncritically or resisting technological change. Through analysis of emerging leadership models across industries—from healthcare executives navigating AI-assisted diagnosis to financial services leaders managing algorithmic trading systems—we explore the competencies, mindsets, and organizational structures required for effective leadership in hybrid intelligence environments. The focus remains practical: what must leaders know, what skills must they develop, and what cultural transformations must they champion to guide organizations through this transition responsibly?

The transformation requires integrating rather than abandoning traditional leadership virtues. Vision, empathy, ethical judgment, and the ability to inspire human teams remain essential—perhaps more so when algorithmic efficiency might otherwise dominate organizational priorities. Yet these enduring capabilities must now combine with new skills such as algorithmic literacy that enables informed oversight without requiring technical expertise, collaborative decision-making approaches that synthesize human and machine insights effectively, and the courage to assert human values when data points toward efficiency gains that conflict with organizational purpose or societal responsibility.

For executives navigating digital transformation, policymakers shaping AI governance frameworks, and professionals across industries preparing for leadership roles in an AI-augmented future, understanding this evolution is not optional. The leaders who will succeed in the coming decades are those who recognize that human judgment becomes more valuable, not less, as AI capabilities expand—but only if that judgment evolves to meet the unique challenges of leading in a world where both humans and machines shape organizational intelligence and strategic direction.

The Leadership Transformation: How AI Changes What Leaders Must Know and Do

The transformation of leadership in the AI era begins with a fundamental shift in knowledge requirements. Leaders no longer have the luxury of delegating AI decisions entirely to technical teams while remaining strategically disengaged.[63] Research from *Harvard Business Review* reveals that many executives have inadvertently ceded their role in the AI transformation by pushing strategy and investment decisions to technology groups that lack comprehensive understanding of organizational goals and values.[63] This abdication creates what researchers describe as a risk of a generation of AI transformation failures unless leaders reconnect AI initiatives to broader strategic imperatives and accountability structures.[63]

Effective leadership now requires what experts term algorithmic literacy—not the ability to code or build models, but rather conceptual fluency in how AI systems function, what they can and cannot accomplish, and where their limitations create risk.[61 62] This literacy enables leaders to set realistic expectations, ask informed questions of technical teams, and recognize when algorithmic recommendations require human override. Understanding that machine learning models depend entirely on training data quality, that correlation does not equal causation, and that optimization for narrow metrics can produce unintended consequences represents essential knowledge for any executive overseeing AI deployment.

Beyond technical understanding, leaders must develop new competencies in human-machine collaboration that fundamentally differ from traditional management skills. Decision-making workflows must be redesigned to specify where AI generates insights, where humans set objectives and constraints, and where final judgment authority resides. Geoff Woods, writing on AI-enabled leadership, describes this as moving from "operational overwhelm to strategic orchestration," using AI to handle computation-heavy analysis while leaders focus attention on long-horizon decisions that require contextual judgment, ethical reasoning, and stakeholder consideration that algorithms cannot provide.[60 61 64]

This orchestration role demands that leaders actively challenge rather than simply accept AI outputs. Woods emphasizes using AI explicitly to interrogate assumptions and surface blind spots by instructing systems to argue against prevailing strategies, highlight disconfirming evidence, and expose confirmation biases that human teams naturally develop.[60] When a leadership team invests heavily in a strategic direction, AI tools can be prompted to analyze data that contradicts that approach, forcing uncomfortable but necessary reconsideration before resources are committed.[60] This requires leaders to view AI not as an authority that delivers answers but as a thought partner that expands the decision space and reveals options that might otherwise remain invisible.

The transformation extends to how leaders communicate and create meaning within organizations. Brian Evergreen, researching leadership in digital transformation contexts, argues that "your strategy and your vision will only be as effective as your ability to articulate them and spark action."[65] As AI automates more operational and analytical work, the distinctly human leadership capacities (vision-setting, narrative-building, ethical judgment, and motivation) become more rather than less critical. Leaders must explain not just what AI will do but why the organization is adopting it, how it aligns with institutional values, and what it means for employees' roles and development.[65]

This requires leaders to act as stewards rather than controllers of AI systems. Stewardship involves establishing governance frameworks that define acceptable uses, required safeguards, and accountability mechanisms before deployment.[63] It means treating data and models as shared organizational assets carrying significant risk, not merely technical tools. Most importantly, it demands that leaders maintain responsibility for outcomes even when AI systems generate the recommendations that inform decisions—recognizing that algorithmic assistance does not diminish human accountability but rather transforms how that accountability must be exercised in increasingly complex socio-technical environments.[63]

Algorithmic Literacy for Executives: Understanding AI Capabilities and Limitations Without Technical Expertise

Executives navigating AI adoption face a paradox: they must govern systems they did not build and cannot fully explain yet remain accountable for the decisions those systems influence. As mentioned previously, this challenge relies heavily on algorithmic literacy.[66 67 68] Gartner research indicates that organizations emphasizing AI literacy for executives will achieve twenty percent higher financial performance by 2027 compared to those that neglect this capability, underscoring that algorithmic understanding has become a component of strategic competence rather than optional technical knowledge.[66]

Algorithmic literacy begins with grasping what AI systems actually do in business contexts. Most practical applications fall into recognizable categories: prediction systems that forecast demand, churn, or delays; classification systems that label transactions as fraudulent or legitimate; recommendation engines that suggest products, prices, or interventions; generative systems that create text, images, or synthetic data; and optimization systems that allocate resources or balance competing objectives.[70 71] Understanding these categories enables leaders to map which decisions AI touches in their enterprise and whether those decisions are operational, tactical, or strategic—a distinction that determines appropriate governance intensity and human oversight requirements.[70 71]

The relationship between AI performance and data quality represents perhaps the most critical concept for non-technical executives to internalize.[67 68 69] Algorithms depend entirely on training data, and biased, incomplete, or poor-quality inputs will propagate into unreliable outputs regardless of model sophistication.[68 69] This reality shifts the executive question from Which algorithm should we use? To What data is this system trained on, who controls it, and how is it audited?[66 67]

Research on AI in professional work demonstrates that domain expertise and algorithmic outputs are complements rather than substitutes and the greatest value emerges when expert judgment and machine recommendations combine thoughtfully, with humans understanding when and how to override AI suggestions based on contextual knowledge that data alone cannot capture.[69]

Executives must also recognize typical AI limitations that create risk when misunderstood. Machine learning models are trained for specific tasks and contexts, therefore using them outside those conditions—what researchers call distribution shift—degrades reliability unpredictably.[66 68 69] Many models detect correlations without explaining causation, creating danger when leaders act as if patterns represent causal relationships suitable for strategic intervention.[69] Complex models may function as black boxes that resist interpretation, a characteristic that matters profoundly in regulated decisions involving credit, employment, or healthcare where transparency requirements exist.[66 67] Historical data can encode discriminatory patterns, leading to unequal treatment across demographic groups.[66 67 68] Perhaps most insidiously, human decision-makers can defer uncritically to AI recommendations, especially when presented with quantitative confidence scores—a phenomenon researchers term automation bias.[68 69]

Algorithmic literacy operationalizes as the ability to ask informed questions rather than supply technical answers.[66 67 68 70 71] For any significant AI initiative, executives should lead conversations addressing five domains: the business problem and strategic fit;[67 70] data sources, quality, and validation methods;[67 68 69] risks to affected stakeholders including bias and compliance concerns;[66 67 68] explainability and accountability mechanisms;[66 67] and lifecycle governance including monitoring, auditing, and controls against unauthorized shadow AI adoption.[66 68] This question framework requires no coding ability but demands that leaders own the consequences of AI-mediated decisions.

The shift toward algorithmic literacy represents a fundamental expansion of leadership responsibility. As AI becomes woven into dashboards, planning systems, and scenario analysis tools, executives increasingly operate in hybrid human-machine decision loops where their role evolves from approving static reports to stewarding decision architectures.[70 71] Leaders must understand when to delegate decisions to AI, when to augment human judgment with algorithmic insights, and when to retain purely human control—distinctions that algorithmic literacy makes visible and governable.

Collaborative Intelligence: Leading Teams Where Both Humans and Machines Contribute Strategic Insights

The fundamental challenge of collaborative intelligence lies not in choosing between human judgment and algorithmic analysis, but in designing decision architectures where each contributes what it does best. Research from MIT's Center for Collective Intelligence, analyzing over one hundred studies of human-AI collaboration, reveals a counterintuitive finding that combining humans and machines does not automatically improve outcomes compared to either working alone.[72] The key determinant of success is whether organizations deliberately structure workflows so that humans handle tasks requiring contextual understanding and ethical reasoning while AI manages pattern detection, simulation, and high-volume data processing.[54 72] When this division of labor aligns with actual capabilities rather than assumptions, human-AI teams can achieve what researchers term synergy with performance exceeding both human-only and AI-only approaches.[72]

This synergy emerges most reliably in domains where tasks can be decomposed into complementary subtasks.[72] In studies of bird species classification, human experts achieved eighty-one percent accuracy while AI systems reached seventy-three percent. When experts collaborated with AI, accuracy jumped to ninety percent because humans demonstrated superior judgment about when to trust algorithmic suggestions and when their own expertise should prevail.[72] Conversely, in fake hotel review detection, AI alone achieved seventy-three percent accuracy compared to fifty-five percent for humans. When combined, performance dropped to sixty-nine percent because humans could not reliably determine when to follow the algorithm, introducing errors through misplaced confidence in their own judgment.[72] These empirical findings underscore a critical leadership principle: effective collaboration requires honest assessment of relative strengths rather than reflexive assumptions that human oversight always adds value.

The implications for strategic leadership extend beyond task allocation to fundamental questions about decision process design.[72] Leaders must specify not just who performs which activities, but how insights flow between human and machine contributors.[54] [72] In strategic communication, for example, AI systems can draft presentations by analyzing audience data, prior engagement metrics, and topical trends, thereby accelerating preparation and expanding the range of options considered.[54] [72] Human leaders then adapt these drafts by infusing narrative coherence, ethical framing, and contextual nuance that algorithms cannot provide.[54] This iterative loop exemplifies collaborative intelligence where AI handles computational intensity while humans ensure meaning and moral resonance.[54] [72]

The organizational architecture supporting such collaboration demands new hybrid roles that embed human judgment throughout AI system lifecycles. Research on human-machine operating models identifies three critical functions: trainers who teach AI systems through data labeling and prompt engineering, explainers who interpret algorithmic recommendations for non-technical stakeholders, and sustainers who monitor systems over time for performance degradation, bias emergence, and value misalignment.[73] These roles ensure that human perspective remains active rather than passive, transforming leaders from approvers of AI output into stewards of socio-technical systems where accountability cannot be delegated to algorithms.

Yet collaborative intelligence carries inherent risks that leaders must actively counter. Over-reliance on algorithmic recommendations can produce what researchers term "moral deskilling"—the erosion of human capacity for nuanced judgment through disuse.[54] When leaders routinely accept AI-generated insights without critical examination, organizations lose institutional memory, narrative understanding, and the ability to recognize when technical optimization conflicts with human values.[54] Strategic decisions become overly data-driven and impersonal, lacking the qualities that inspire commitment and build legitimacy with stakeholders.[54] Preventing this degradation requires deliberate practices: structured forums where teams review not just what AI recommended but why they accepted or rejected it, explicit encouragement of dissent from algorithmic outputs, and clear boundaries around decisions that must remain under human control regardless of AI capability.

The path forward demands that leaders view AI neither as authority delivering answers nor as mere tool executing commands, but as thought partner expanding the decision space while humans retain ultimate responsibility for choices and their consequences.[54]

The Judgment Imperative: Cultivating Executive Discernment About When to Trust or Override AI Recommendations

The most critical leadership capability emerging from human-AI collaboration is executive discernment, the ability to determine when algorithmic recommendations warrant trust and when they demand override. This imperative judgment represents neither blanket acceptance of machine intelligence nor reflexive resistance to it, but rather a disciplined, context-sensitive approach to evaluating AI outputs against organizational values, domain expertise, and situational factors that algorithms cannot fully capture. Research from decision intelligence frameworks reveals that effective executives develop systematic criteria for this evaluation rather than relying on intuition alone, recognizing that both automation bias (excessive trust in algorithmic recommendations) and algorithm aversion (systematic distrust following a single failure) can undermine decision quality.

The judgment imperative begins with classifying decisions along dimensions that determine appropriate trust levels. High-stakes, irreversible decisions with significant ethical implications such as workforce reductions, strategic pivots affecting communities, or resource allocations with safety consequences demand heightened human scrutiny regardless of algorithmic confidence. Conversely, low-stakes, reversible decisions in time-sensitive contexts may warrant greater deference to AI recommendations, particularly when human fatigue or cognitive load compromises judgment quality. A study analyzing human-AI collaboration across multiple domains found that performance improvements emerged not from always trusting or always overriding AI, but from matching the collaboration model to decision characteristics, with human-in-the-loop oversight for consequential choices and human-over-the-loop monitoring for high-volume operational decisions.

Executives must also evaluate the AI system's reliability within the specific decision context. Algorithms trained on historical data may fail, leading to distribution shift and making performance metrics from training environments unreliable predictors of real-world accuracy. Leaders should question whether the current situation resembles the scenarios on which the system was trained, whether the data reflects relevant patterns or merely historical artifacts, and whether the AI can quantify its own uncertainty. Systems that express confidence levels enable more nuanced human oversight, allowing leaders to intervene when algorithmic certainty is low while deferring when confidence is justified by validation evidence.

The transparency and explainability of AI recommendations fundamentally affect whether override is feasible. When systems provide interpretable reasoning—identifying which factors drove a recommendation and how alternative inputs would change outputs—domain experts can assess whether the algorithmic logic aligns with causal understanding and ethical principles. A healthcare executive reviewing an AI-generated treatment protocol can evaluate whether the system appropriately weighted patient-specific factors, comorbidities, and quality-of-life considerations that clinical judgment would prioritize. Conversely, opaque recommendations from black-box models leave leaders unable to distinguish between sophisticated pattern recognition and spurious correlations, making informed override decisions nearly impossible.

Organizations that excel at the judgment imperative establish explicit governance protocols defining which decision types require human review, which can be automated with monitoring, and which demand executive approval. These protocols create what researchers describe as "decision workflows" that specify not just who decides but how human and machine contributions integrate. Financial services firms, for example, may automate routine fraud detection while requiring human review for cases above certain thresholds or involving vulnerable populations, with escalation paths to senior leadership when AI recommendations conflict with regulatory principles or reputational considerations.

Perhaps most critically, effective executives recognize that the judgment imperative is not a one-time assessment but an ongoing stewardship responsibility. They implement feedback mechanisms that track when humans override AI, analyze the outcomes of those overrides, and use this data to refine both algorithmic models and human decision criteria. This continuous learning approach treats human-AI collaboration as an evolving capability requiring deliberate cultivation rather than a static technical implementation, ensuring that organizational discernment improves as both human understanding and algorithmic sophistication advance.

Building AI-Ready Organizational Cultures: Fostering Environments That Balance Innovation with Human-Centered Values

The most sophisticated AI systems will fail to deliver value—and may cause significant harm—if deployed into organizational cultures unprepared to use them responsibly. Research from MIT's Center for Collective Intelligence analyzing over one hundred studies of human-AI collaboration reveals a sobering finding that combining humans and machines does not automatically improve outcomes compared to either working alone.[72] The determining factor is whether organizations deliberately structure workflows, norms, and decision processes so that each party contributes what it does best. As Thomas Malone summarizes, "Combinations of humans and AI work best when each party can do the thing they do better than the other."[72] This insight underscores that building AI-ready organizational cultures is not about maximizing technology adoption but about fostering environments where innovation serves rather than overrides human-centered values.

An AI-ready culture exhibits several distinguishing characteristics that extend far beyond technical infrastructure. Strategic clarity about AI's role ensures the technology is framed as augmenting human capabilities rather than replacing human agency.[73] Ethical orientation means deployment decisions are guided by explicit values including fairness, accountability, transparency, and dignity.[54] A learning orientation treats AI as a catalyst for continuous improvement through pilots, feedback loops, and iterative refinement rather than one-time automation projects.[72] Psychological safety enables employees to question algorithmic outputs, report harms, and raise ethical concerns without fear of retaliation.[54] Finally, widespread algorithmic literacy gives non-technical staff sufficient understanding to interpret AI recommendations, spot anomalies, and recognize when human override is appropriate.[72]

These cultural attributes matter because they address fundamental tensions inherent in AI adoption. Organizations face constant pressure to choose between efficiency and empathy. AI excels at optimizing for speed and cost but can produce decisions that are technically optimal yet ethically questionable or emotionally tone-deaf.[54] The tension between scale and personalization emerges as algorithms process millions of cases efficiently while potentially neglecting local context, cultural nuance, or individual circumstances.[54] Perhaps most critically, over-reliance on AI risks eroding experiential learning and reflective judgment as humans defer to algorithmic outputs rather than deeply engaging with decisions.[54] AI-ready cultures make these tensions explicit and embed systematic counterweights like human review protocols, ethics committees, and appeal mechanisms, therefore ensuring innovation is balanced with human-centered values rather than allowed to dominate uncritically.[54]

Leadership in such cultures shifts from command-and-control to orchestrating human-machine systems. Leaders must engage in sensemaking, helping teams understand how AI tools work, where their strengths and limits lie, and how they connect to organizational mission. Boundary setting defines "no-go zones" for automation—decisions that must retain human control regardless of technical capability—and establishes minimum oversight levels for different decision types. Capability building invests in human skills complementary to AI, including contextual reasoning, empathy, domain expertise, and cross-disciplinary collaboration.[72] Ethical stewardship champions transparency, fairness, and accountability in AI projects, including willingness to halt or redesign systems that conflict with values.

The MIT research emphasizes that redefining processes proves more effective than simply reassigning tasks between humans and AI.[72] Rather than asking which human tasks can be automated, AI-ready leaders map end-to-end decision processes, identify subtasks where AI excels versus where humans excel, and design hybrid workflows where AI supports human decision-makers with recommendations and scenario analyses while humans retain final accountability in high-stakes decisions.[72] This approach treats AI deployment as socio-technical intervention requiring continuous monitoring of both performance metrics and long-term effects on people—including skill development, cognitive load, interpersonal dynamics, and organizational learning capacity. The result is not maximum automation but optimized collaboration where technology amplifies rather than diminishes human judgment, creativity, and moral reasoning.

Leadership Competencies for the Hybrid Future: Integrating Traditional Virtues with New Skills for the AI Age

The integration of traditional leadership virtues with AI-era competencies represents neither a wholesale replacement of timeless qualities nor a simple addition of technical skills, but rather a fundamental reconceptualization of what leadership demands when both humans and machines contribute to organizational intelligence.[75] [62] [78] Research from MIT's Initiative on the Digital Economy demonstrates that the most effective leaders in AI-augmented environments exhibit what scholars term a "moral-technical integration"—the capacity to simultaneously uphold enduring ethical principles while developing fluency in algorithmic systems and their governance implications.[75] [62] [78] [79] This integration matters because AI deployment decisions are never purely technical; they embody choices about organizational values, stakeholder priorities, and the kind of future leaders wish to create.[62]

Traditional virtues take on new dimensions in hybrid human-AI contexts. Practical wisdom, long recognized as the ability to discern right action in complex situations, now requires leaders to contextualize algorithmic recommendations within broader strategic and ethical frameworks that machines cannot access.[74] [62] [78] When an AI system proposes workforce restructuring based on efficiency optimization, practical wisdom enables leaders to weigh factors the algorithm cannot quantify such as community commitments, institutional knowledge, long-term reputation effects, and human dignity.[62] [78] Justice and fairness, foundational to ethical leadership across centuries, demand heightened attention when algorithmic systems can encode and amplify historical biases present in training data.[74] [62] [78] Leaders committed to justice must ensure that AI-assisted decisions affecting hiring, credit, healthcare, and resource allocation are scrutinized for disparate impacts across demographic groups, treating fairness as a design constraint rather than a post-deployment correction.[62 78]

Courage becomes of particular importance when leaders must resist pressures to automate decisions simply because technology makes it possible or competitors are doing so.[62 78] Moral courage manifests in the willingness to pause profitable AI deployments that conflict with organizational values, to challenge assumptions that efficiency should always trump other considerations, and to make unpopular investments in safeguards that slow implementation but protect stakeholders.[62 78] Integrity—consistency between stated values and actions—becomes highly visible in AI adoption, as stakeholders scrutinize whether organizations' algorithmic practices align with their public commitments regarding privacy, transparency, and social responsibility.[75 76 62] Humility takes an epistemic form, requiring leaders to acknowledge both their own limitations in understanding complex AI systems and the systems' limitations in capturing contextual nuances and ethical considerations that humans recognize intuitively.[74 62]

These enduring virtues must combine with new competencies specific to the AI age. Algorithmic literacy enables leaders to ask informed questions about data provenance, model objectives, performance metrics, and testing protocols without requiring technical expertise in machine learning.[74 62 78] This literacy transforms executives from passive consumers of AI recommendations into active stewards capable of meaningful oversight.[62 78 79] Ethical foresight—the capacity to anticipate second- and third-order consequences of AI deployment—allows leaders to identify value conflicts early, before systems are embedded in organizational processes and difficult to modify.[62 78] Human-centered design competency ensures that leaders prioritize dignity, agency, and experience in AI-enabled workflows, creating systems where humans can understand, contest, and override algorithmic decisions when appropriate.[62 78]

Collaborative intelligence represents perhaps the most distinctive new leadership skill: the ability to orchestrate hybrid teams where humans and AI systems each contribute what they do best.[74 62 78] This requires designing decision workflows that specify when AI handles pattern detection and scenario analysis, when humans apply contextual judgment and ethical reasoning, and how insights flow between machine and human contributors.[74 75 62 78] Leaders who excel at collaborative intelligence treat AI neither as authority delivering answers nor as mere tool executing commands, but as thought partner expanding the decision space while humans retain ultimate accountability for choices and their consequences.[74 62 77 78] This integration of traditional virtues with emerging competencies defines leadership adequacy for organizations navigating the transition to AI-augmented decision-making.[75 76 62 78 79]The transformation of leadership in the age of AI-assisted decision-making represents neither a technical upgrade to existing management practices nor a wholesale abandonment of traditional executive virtues, but rather a fundamental reconceptualization of what it means to guide organizations when both humans and machines contribute to strategic intelligence. The scenarios explored throughout this chapter—from executives weighing algorithmic efficiency recommendations against community commitments to leaders orchestrating hybrid teams where judgment flows between human and machine contributors—reveal that successful navigation of this transition demands integration rather than choice. Leaders cannot simply delegate AI governance to technical teams, nor can they resist algorithmic capabilities that competitors are leveraging. Instead, they must develop new competencies while deepening rather than discarding the distinctly human qualities that have always defined effective leadership.

The evidence presented demonstrates that algorithmic literacy, collaborative intelligence, and ethical discernment are not optional enhancements but core leadership requirements for organizations deploying AI decision-making systems. Executives who lack conceptual understanding of how AI functions, what it can accomplish, and where its limitations create risk cannot fulfill their accountability obligations regardless of technical staff expertise. Research showing that organizations emphasizing AI literacy for executives achieve twenty percent higher financial performance underscores that this knowledge directly affects competitive outcomes, not merely governance compliance. Similarly, the ability to design decision workflows where humans and machines each contribute what they do best, rather than assuming that combining them automatically improves results, separates leaders who capture AI's value from those who experience expensive implementation failures.

Yet the chapter's most critical insight concerns not what leaders must learn but what they must preserve. The judgment imperative—knowing when to trust algorithmic recommendations and when human override is essential—cannot be reduced to technical criteria or decision matrices. It requires practical wisdom, moral courage, and contextual understanding that algorithms cannot replicate. When AI systems optimize for efficiency in ways that conflict with organizational values, community commitments, or stakeholder dignity, leaders must possess both the discernment to recognize the conflict and the courage to assert human judgment over machine recommendations. This capacity distinguishes leadership from management, strategy from optimization, and organizations that use AI responsibly from those that allow algorithmic logic to dominate decision-making uncritically.

The cultural transformation required extends beyond individual leader development to encompass organizational norms, decision processes, and accountability structures. AI-ready cultures exhibit strategic clarity about technology's role, ethical orientation in deployment decisions, psychological safety for questioning algorithmic outputs, and widespread literacy that enables non-technical employees to interpret recommendations and recognize when human intervention is appropriate. Building such cultures demands that leaders act as stewards rather than controllers by establishing governance frameworks, defining boundaries around decisions that must remain under human control, and investing in capabilities complementary to AI including contextual reasoning, empathy, and cross-disciplinary collaboration.

For executives navigating digital transformation, the path forward requires neither embracing automation uncritically nor resisting technological change defensively. It demands recognizing that human judgment becomes more valuable, not less, as AI capabilities expand—but only if that judgment evolves to meet the unique challenges of leading in hybrid human-machine environments. The leaders who will guide organizations successfully through this transition are those who understand that AI is neither authority delivering answers nor mere tool executing commands, but collaborative partner that expands decision space while humans retain ultimate responsibility for choices and their consequences. This integration of traditional virtues with emerging competencies defines leadership adequacy for the AI age, transforming abstract questions about algorithmic governance into practical frameworks that professionals can implement immediately as intelligent systems become increasingly embedded in organizational decision-making.

CHAPTER 6.

THE AI TRANSFORMATION ROADMAP: PRACTICAL STEPS FOR RESPONSIBLE ADOPTION

A pharmaceutical company's executive team approved an ambitious AI transformation initiative with a clear timeline, substantial budget, and enthusiastic board support, yet eighteen months later the project had stalled—not due to technical failures, but because no one had assessed whether the organization's culture, governance structures, and employee capabilities were ready for the shift. The expensive algorithms sat underutilized while employees circumvented the new systems, reverting to familiar processes because the transformation roadmap had focused exclusively on technology deployment while neglecting the human, organizational, and ethical infrastructure required for successful adoption.

This scenario repeats itself across industries with troubling frequency. Organizations invest millions in sophisticated AI capabilities, hire talented data scientists, and secure executive sponsorship, yet struggle to translate technological potential into responsible operational reality. The gap between AI aspiration and implementation success reveals a fundamental misunderstanding. Adopting AI decision-making systems is not primarily a technology challenge but a transformation challenge that touches every aspect of organizational functioning.

The previous chapters have explored the theoretical foundations, ethical frameworks, and leadership competencies required for responsible AI adoption. This chapter shifts from conceptual understanding to practical execution, providing a structured roadmap that organizations can follow to implement AI decision-making systems while maintaining human accountability and ethical integrity. The framework presented here synthesizes lessons from successful transformations across industries, identifying the critical steps that separate initiatives that deliver sustainable value from those that stall, fail, or create unintended consequences.

The transformation journey begins not with technology selection but with honest organizational assessment. Before deploying any AI system, leaders must evaluate whether their culture encourages questioning algorithmic recommendations, whether governance structures can assign clear accountability for machine-generated decisions, and whether employees possess the capabilities to collaborate effectively with intelligent systems. Organizations that skip this foundational assessment inevitably encounter resistance, confusion about responsibility, and systems that operate in isolation from the human judgment they were meant to augment.

Equally critical is establishing governance frameworks before deployment rather than retrofitting accountability mechanisms after problems emerge. The most successful AI implementations create clear lines of responsibility, define decision rights between humans and machines, and establish monitoring protocols before the first algorithm goes live. This governance-first approach prevents the accountability vacuums that have plagued organizations where AI systems make consequential decisions without anyone clearly responsible for outcomes.

The roadmap also emphasizes the strategic value of pilot programs that test not just technical performance but ethical implications and organizational readiness. Small-scale implementations allow organizations to identify unforeseen challenges, refine governance processes, and build employee confidence before scaling systems across the enterprise. These pilots serve as learning laboratories where companies discover what responsible AI adoption actually requires in their specific context rather than abstract theory.

As organizations move from pilot to enterprise-wide deployment, maintaining human oversight becomes increasingly challenging. The framework addresses this scaling challenge directly, providing strategies for preserving meaningful human agency even as AI systems handle growing volumes of decisions. This includes designing intervention points where human judgment can override algorithmic recommendations, creating feedback loops that capture when systems produce questionable outputs, and building organizational capabilities for ongoing AI stewardship rather than one-time implementation.

The chapter recognizes that responsible AI adoption is not a destination but a continuous journey. Technology evolves, organizational contexts shift, and societal expectations change, requiring adaptive governance and sustained commitment to ethical principles. The roadmap therefore concludes with frameworks for continuous monitoring, assessment, and course correction—transforming AI transformation from a project with a defined endpoint into an ongoing organizational capability.

For executives, consultants, and policymakers navigating the practical realities of AI adoption, this chapter offers concrete tools and decision criteria that can be immediately applied. The goal is not to provide a universal template that fits every organization, but rather a flexible framework that can be adapted to different industries, organizational cultures, and risk tolerances while maintaining core principles of accountability, transparency, and human-centered design.

Organizational Readiness Assessment: Evaluating Cultural, Structural, and Capability Prerequisites Before AI Adoption

Before committing resources to AI implementation, organizations must conduct a rigorous assessment across three interdependent dimensions: cultural readiness, structural capacity, and organizational capabilities. This evaluation reveals whether the institution possesses the prerequisites for responsible adoption or whether foundational work is required before deployment can succeed.[81] [82] Research on digital transformation consistently shows that organizations investing first in readiness assessment experience fewer implementation failures and compliance issues than those rushing directly to technology deployment.[81]

Cultural readiness examines whether organizational norms, leadership behaviors, and employee attitudes can support responsible AI adoption. The assessment begins with leadership mindset—whether executives articulate a clear vision connecting AI to organizational purpose and ethical commitments, not merely efficiency gains.[81] [82] Leaders who model responsible AI use, openly discuss ethical constraints, and frame AI as capability amplification rather than workforce replacement create environments where adoption succeeds.[80] [81] Equally critical is psychological safety. Employees must feel secure questioning algorithmic outputs, experimenting with new tools, and raising concerns about potential harms without fear of reprisal.[80] Organizations where AI is perceived primarily as surveillance or job elimination face resistance that undermines even technically sound implementations.[80]

The cultural assessment also evaluates the organization's ethical climate—whether shared values regarding fairness, transparency, and accountability exist before AI introduces new decision-making complexity. Mature AI-ready cultures create space for ethical deliberation through mechanisms like cross-functional review committees and encourage employees to speak up when systems produce questionable results.[81] This ethical infrastructure cannot be retrofitted after deployment; it must exist as organizational foundation.[81]

Structural readiness focuses on governance frameworks, organizational design, and formal processes that assign accountability and enable oversight. The assessment examines whether the organization has designated executive sponsorship for AI initiatives, typically through roles like Chief Data and AI Officer, and whether cross-functional governance bodies exist to oversee priorities, risks, and compliance.[81 82] Critical questions include whether policies document expectations for data use, model lifecycle management, human oversight requirements, and monitoring protocols.[81] Organizations lacking these formal structures inevitably encounter accountability vacuums when AI systems produce harmful outcomes.

The structural assessment also evaluates organizational design choices—whether AI capabilities reside in centralized teams, distributed across business units, or operate through hub-and-spoke models. [81] Each approach carries implications for knowledge sharing, quality control, and scalability. Successful organizations establish mechanisms like communities of practice and rotating AI champions to diffuse expertise across departments rather than concentrating knowledge in isolated technical teams.[80] Process maturity matters equally. Documented workflows for use-case prioritization, model validation, ongoing monitoring, and incident response distinguish organizations prepared for responsible scaling from those operating reactively.[81]

Capability readiness addresses the practical resources required for AI adoption: data maturity, technical infrastructure, human skills, and change management capacity. Data readiness consistently emerges as the primary constraint on AI success as organizations must assess whether critical datasets are accessible, high-quality, well-documented, and governed by clear policies on ownership, consent, and security.[81] Technical infrastructure evaluation examines whether scalable compute, storage, and deployment environments exist and can integrate with legacy systems.[81 82]

Human capability assessment extends beyond technical talent to include domain expertise, ethical and legal specialists, and broad workforce AI literacy.[81] Organizations need data scientists and engineers capable of building systems, business leaders who can frame problems correctly and interpret outputs, and ethics experts who understand AI-specific issues like bias and explainability.[81 82] Equally important is change management capacity—the ability to manage behavioral and process transformation through structured communication, training, feedback loops, and adjusted incentive systems.[80 81]

Organizations that conduct comprehensive readiness assessments before AI adoption identify gaps that would otherwise derail implementation, align stakeholders around shared understanding of prerequisites, and establish baselines for measuring transformation progress.[81 82] This diagnostic work transforms AI adoption from technology deployment into organizational transformation.

Governance-First Implementation: Establishing Accountability Frameworks and Ethical Guidelines Before Deployment

The most consequential governance decision organizations make about AI systems occurs before the first line of code is written: establishing who will be accountable when things go wrong. A governance-first approach inverts the typical technology adoption sequence, requiring that accountability frameworks, ethical guidelines, and oversight mechanisms be designed and agreed upon before deployment rather than retrofitted after problems emerge. This foundational shift transforms governance from a compliance afterthought into a design constraint that shapes what gets built and how it operates.

Organizations that deploy AI without prior governance structures predictably encounter a common set of failures: data breaches, discriminatory outcomes, regulatory violations, and accountability vacuums where no individual or team can explain why a system made a particular decision.[17] Research on AI governance consistently demonstrates that these failures stem not from technical deficiencies but from institutional gaps—the absence of clear decision rights, oversight mechanisms, and escalation pathways before systems go live.[17 18] The governance-first approach addresses this by making institutional architecture a prerequisite rather than a reaction.

The foundation of any accountability framework is structural clarity about roles and responsibilities. Effective governance begins at the board level, where a designated committee—whether focused on risk, audit, technology, or ethics—assumes explicit responsibility for AI oversight analogous to how boards govern financial risk or cybersecurity.[18 19] This committee requires regular reporting on AI risk posture, significant incidents, and mitigation progress, ensuring that AI governance receives the same executive attention as other material risks to the organization.

Below the board, organizations must designate executive ownership through a senior leader accountable for AI strategy, compliance, and risk management.[17 18] Whether this responsibility resides with a Chief AI Officer, Chief Data Officer, or shared governance role matters less than ensuring one individual has clear authority and accountability. This executive then establishes a cross-functional AI governance council that includes representatives from engineering, legal, compliance, information security, human resources, and affected business units.[17 18]

This council's mandate extends beyond advisory input to actual decision authority: approving AI use cases before development, setting technical standards and ethical policies, and reviewing high-risk models before launch.[17 19]

For each significant AI system, the framework assigns three distinct accountability roles: a business owner responsible for outcomes and ethical impact, a technical owner responsible for design and model behavior, and a risk owner responsible for policy adherence and regulatory compliance.[17 19] This allocation operationalizes the principle that AI governance cannot be separated from overall corporate governance, but that it must be integrated into existing management structures with the same rigor applied to other consequential business decisions.[83]

Governance-first implementation also requires translating abstract ethical principles into enforceable operational policies before development begins. Many organizations adopt high-level commitments to fairness, transparency, and accountability, yet these remain aspirational without concrete translation into workflow requirements.[83] [19] Effective frameworks specify, for example, that fairness principles mandate documented protocols for bias assessment, defined thresholds for acceptable outcome disparities across demographic groups, and clear remediation procedures when systems exceed those thresholds.[17 18] Transparency commitments translate into requirements for model documentation, decision logging, and user-facing explanations of how systems reach conclusions.[17 18]

Critically, these policies function as workflow constraints rather than static documents.[17] Risk assessments, approval gates, and testing protocols are embedded into development pipelines, making governance part of how work gets done rather than a separate audit function.[17] This integration ensures that ethical considerations shape technical choices from the outset and influence model selection, data collection practices, and deployment architectures before systems are built.

The governance-first approach recognizes that accountability cannot be retrofitted after deployment. By establishing clear responsibility structures and operational policies before the first algorithm goes live, organizations create the institutional foundation required for responsible AI adoption at scale.

The Pilot Program Strategy: Designing Small-Scale Tests That Evaluate Both Performance and Ethical Implications

Once governance structures are established, organizations face the critical question of how to test AI systems before enterprise-wide deployment. Pilot programs serve as controlled experiments that evaluate not only whether algorithms perform as designed but whether they operate ethically within real organizational contexts.[85] These small-scale implementations function as learning laboratories where companies discover what responsible AI adoption actually requires in their specific environment rather than abstract theory.

The fundamental principle distinguishing ethical pilots from conventional technology tests is dual evaluation criteria. Traditional pilots measure technical performance like accuracy rates, processing speed, system reliability, and integration compatibility. Ethical pilots add a parallel assessment track examining fairness across demographic groups, explainability of decisions, privacy protection, human oversight quality, and alignment with organizational values.[84 85] Research on AI risk management emphasizes that organizations must proactively identify and mitigate harms around bias, explainability, and privacy before scaling systems, treating ethics as a risk management challenge rather than compliance afterthought.[85]

Designing an effective pilot begins with selecting use cases that balance organizational value against controlled risk exposure. The ideal pilot addresses a decision domain important enough to reveal real trade-offs and ethical tensions, yet not so safety-critical that initial errors could produce catastrophic consequences. Organizations should choose contexts with well-documented baseline decision-making processes, enabling robust comparison between human and AI-assisted outcomes. Equally important is clear understanding of which populations could be disproportionately affected—patients at different triage levels, loan applicants across demographic categories, employees in various roles—allowing targeted fairness assessment.[43 85]

The pilot scope must explicitly define decision authority boundaries.[84] Organizations should document whether the AI functions as a recommendation system supporting human judgment or an autonomous decision-maker, what maximum decision severity the system may handle during testing, and under what conditions humans must override algorithmic outputs.[84] This clarity prevents the accountability confusion that emerges when roles remain ambiguous during deployment.

Ethical impact assessment requires transforming abstract concerns into testable hypotheses with measurable outcomes. Rather than vaguely committing to fairness, organizations should articulate specific predictions: "The credit scoring model will not increase approval rate disparities between demographic groups relative to current manual processes" or "Explanations provided will be comprehensible to loan officers and will support meaningful appeal procedures." Each hypothesis shapes data collection protocols and analysis plans.

Fairness evaluation demands measuring system performance disaggregated by relevant subgroups—examining error rates, calibration quality, and treatment rates across protected categories.[85] Organizations must select appropriate fairness definitions for their context, whether demographic parity, equalized odds, or predictive parity, documenting the rationale and acknowledging trade-offs between competing fairness criteria.

Explainability assessment evaluates whether staff can articulate in their own words why the AI recommended specific decisions, whether explanation tools support meaningful challenge mechanisms, and whether consistent procedures exist for handling human-AI disagreements.[85] Privacy protection evaluation examines data minimization practices, anonymization effectiveness, access controls, and compliance with relevant standards.[85] Pilots should test for potential misuse scenarios including function creep, unauthorized secondary use, and re-identification risks through adversarial review processes.

Beyond technical metrics, ethical pilots must assess human experience dimensions: trust in decisions and institutions, perceived procedural fairness, psychological impacts on affected individuals, and changes in worker autonomy and job satisfaction.[84] These qualitative dimensions require surveys, structured interviews, and ethnographic observation alongside quantitative performance data.

The pilot governance structure should include pre-deployment ethical risk review,[85] real-time monitoring with alert thresholds for performance degradation or fairness violations, accessible incident reporting channels for staff and participants, and clear decision rights for pausing or modifying the pilot when ethical boundaries are crossed. This continuous oversight transforms ethics from one-time approval into ongoing stewardship, ensuring that organizations learn what responsible AI implementation demands before committing to irreversible scaling decisions.

Scaling with Safeguards: Moving from Pilot to Enterprise-Wide Deployment While Maintaining Human Oversight

The transition from successful pilot to enterprise-wide deployment represents the moment when organizations confront the fundamental tension between operational efficiency and meaningful human oversight. Pilots succeed precisely because they operate in controlled environments with intensive human attention, limited scope, and flexible intervention mechanisms. Scaling these systems across thousands of daily decisions fundamentally alters this dynamic, creating pressures that can erode the very safeguards that made pilot success possible.[31]

Research on AI governance demonstrates that the psychological costs of oversight increase substantially as decision volume grows. Field studies reveal that human reviewers experience measurable cognitive burden when monitoring high volumes of AI decisions, potentially degrading the quality of oversight mechanisms designed to maintain safety and accountability.[86] This creates a paradox. The more decisions an AI system handles, the more critical human oversight becomes, yet the volume simultaneously increases cognitive load on reviewers, leading to decision fatigue and reduced attention to genuine errors.[86]

Organizations must therefore design scaling strategies that work with rather than against human cognitive limitations. This requires moving beyond the assumption that all AI decisions warrant identical oversight intensity, instead implementing differentiated governance frameworks that match decision types to appropriate human-machine collaboration models.[31]

For high-stakes decisions with significant consequences, organizations should maintain human-in-the-loop governance where AI recommendations require explicit human approval before implementation.[31] This model proves most suitable for decisions affecting individual rights, safety-critical choices, or contexts where regulatory frameworks mandate human review, such as those covered under Article 22 of the European Union's General Data Protection Regulation.[31] Scaling this approach requires limiting its application to genuinely high-stakes, lower-volume decisions where error costs justify review overhead, and establishing specialized review teams with domain expertise rather than generic oversight processes.

For moderate-risk decisions requiring rapid response, human-over-the-loop models allow AI systems to implement decisions that humans can quickly override if necessary.[31] This approach scales more effectively than mandatory pre-approval while preserving intervention capability. Examples include fraud detection systems that automatically disable credit cards when suspicious activity is detected, or semi-autonomous vehicles that brake to avoid collisions.[31] Successful scaling of this model demands robust monitoring systems that alert humans only to decisions requiring review, preventing the alert fatigue that undermines oversight effectiveness.

Certain decision categories warrant granting machines authority over human objections, particularly safety-critical systems where human decision-making may be impaired or where response speed is paramount.[31] Cybersecurity systems that prevent employees from sending malicious attachments or autonomous safety systems that override impaired drivers exemplify this model.[31] Such delegation requires extensive pre-deployment validation, continuous monitoring to ensure system reliability, and transparent communication with stakeholders about why machines have been granted authority in specific domains.

The most sophisticated scaling approach implements hybrid human-machine decisions with escalation protocols.[31] In this model, humans and machines independently evaluate decisions, with discrepancies triggering review by senior decision-makers.[31] This prevents anchoring bias where humans unconsciously defer to machine recommendations, though it demands investment in systems architecture supporting parallel decision-making and clear escalation protocols for high-risk disagreements.[31]

Regardless of governance model selected, enterprise deployment must preserve space for human judgment regarding ethical considerations, contextual factors specific to organizational culture, and stakeholder impacts beyond numerical optimization metrics. As AI systems notoriously fail at capturing intangible human factors—the ethical, moral, and contextual considerations that guide real-world decision-making—scaling strategies must explicitly maintain mechanisms for human moral reasoning rather than pursuing AI autonomy as the ultimate objective.[87]

Organizations that succeed in scaling AI deployment recognize that human oversight is not a temporary constraint to minimize but an essential component of responsible decision-making that must be designed into systems from inception.

Building AI Stewardship Capabilities: Developing Organizational Competencies for Long-Term Responsible Management

Successful scaling of AI systems creates a new organizational challenge: building the sustained competencies required to manage these systems responsibly over time. Unlike traditional technology implementations that stabilize after deployment, AI decision-making systems require ongoing stewardship via continuous monitoring, adaptation, and governance that evolves alongside both technological capabilities and organizational contexts. This level of support demands developing institutional competencies that extend far beyond the technical teams who build AI systems to encompass the entire organization.

The stewardship challenge begins with recognizing that AI systems are not static deployments but dynamic capabilities that change as they learn from new data, as business conditions shift, and as societal expectations evolve. Organizations must therefore build competencies for continuous oversight rather than one-time implementation, transforming AI governance from a project phase into an operational discipline. Research on AI governance frameworks consistently emphasizes that responsible AI management requires sustained organizational attention, not merely initial deployment safeguards.

The foundation of stewardship capability is decision architecture literacy across leadership levels. Executives and managers must develop the competency to recognize which decisions warrant different levels of AI involvement and human oversight. This requires moving beyond simplistic "automate everything possible" or "humans must always decide" mindsets toward nuanced understanding of decision contexts.[31] Leaders need frameworks for evaluating when AI recommendations should require explicit human approval, when machines can implement decisions that humans can override, and when hybrid human-machine processes create superior outcomes to either operating independently.[31]

Building this literacy demands structured education programs that help non-technical leaders understand AI capabilities and limitations without requiring deep technical expertise. Organizations that succeed in developing stewardship capabilities invest in executive education that translates algorithmic concepts into business decision-making contexts, enabling leaders to ask informed questions about system behavior, recognize when AI outputs warrant skepticism, and understand the trade-offs inherent in different governance models.

Equally critical is developing organizational competency in bias recognition and mitigation that extends beyond technical fairness metrics to encompass contextual judgment about acceptable outcomes.[88] While data scientists can measure statistical disparities across demographic groups, determining whether those disparities constitute unacceptable bias requires domain expertise, ethical reasoning, and stakeholder input that technical teams alone cannot provide. Stewardship capability means building cross-functional review processes where business leaders, legal experts, ethicists, and affected community representatives collaborate to evaluate whether AI systems operate fairly within specific organizational and social contexts.

Organizations must also develop competencies for continuous performance monitoring that tracks not just technical accuracy but ethical alignment and stakeholder impact. This requires establishing metrics that measure decision quality across multiple dimensions like efficiency gains, error rates, fairness across subgroups, user trust, employee satisfaction with AI collaboration, and alignment with organizational values. Effective monitoring systems create alert mechanisms that flag not only technical failures but also ethical boundary violations such as when outcome disparities exceed predetermined thresholds or when explanation quality degrades below acceptable standards.

The stewardship competency extends to organizational learning mechanisms that capture lessons from both successes and failures. Organizations should establish structured processes for reviewing AI system performance, documenting incidents where systems produced problematic outcomes, analyzing root causes, and translating those insights into improved governance practices. This institutional memory prevents organizations from repeatedly encountering the same ethical pitfalls and enables continuous refinement of decision architectures as experience accumulates.

Building these capabilities requires sustained investment in people, processes, and governance structures rather than one-time training initiatives. Organizations must create dedicated roles focused on AI stewardship, establish communities of practice that share lessons across business units, and integrate AI governance responsibilities into existing management structures rather than treating them as separate technical functions. The goal is embedding responsible AI management into organizational DNA—making stewardship a core competency rather than a specialized activity confined to technical teams.

Continuous Monitoring and Adaptation: Creating Feedback Loops for Ongoing Assessment and Course Correction

The most sophisticated AI governance frameworks and stewardship capabilities remain incomplete without mechanisms to detect when systems drift from their intended behavior or when organizational contexts shift in ways that render initial design decisions obsolete. Continuous monitoring and adaptation transform AI implementation from a one-time project into an ongoing organizational discipline, creating feedback loops that enable course correction before minor performance degradations escalate into significant failures or ethical breaches.

Research on AI system lifecycle management consistently demonstrates that model performance degrades over time as the statistical patterns learned from historical training data diverge from current operational reality—a phenomenon known as data drift. A credit scoring algorithm trained on pre-pandemic economic patterns, for example, may produce systematically biased predictions when applied to post-pandemic financial behaviors, yet without active monitoring, organizations may continue relying on these degraded outputs for months before recognizing the problem. Continuous monitoring addresses this challenge by establishing systematic observation protocols that track both technical performance metrics and broader indicators of ethical alignment and stakeholder impact.

Effective monitoring architectures operate across multiple dimensions simultaneously. Technical performance tracking measures accuracy rates, processing speeds, error patterns, and system reliability, the quantitative metrics that indicate whether algorithms function as designed. Organizations should establish automated dashboards that flag when key performance indicators fall below predetermined thresholds, triggering investigation before degradation becomes severe. Equally critical is disaggregated performance analysis that examines whether system accuracy varies across demographic groups, geographic regions, or other relevant subpopulations, revealing fairness issues that aggregate metrics might obscure.

Beyond technical metrics, monitoring must capture human experience dimensions that quantitative data alone cannot fully represent. This includes tracking how effectively humans and machines collaborate in hybrid decision contexts, measuring whether human override rates suggest appropriate trust calibration or problematic automation bias, and assessing whether explanation quality enables meaningful human oversight.[31] Organizations should implement structured feedback channels where employees can report concerns about system behavior, creating qualitative data streams that complement quantitative monitoring.

The adaptation component transforms monitoring insights into action through structured decision protocols that determine when and how to adjust AI systems. Organizations must establish clear governance rules specifying what types of performance changes warrant different responses—from minor parameter adjustments that technical teams can implement autonomously to fundamental architecture changes requiring executive approval and stakeholder consultation. This governance clarity prevents both overreaction to normal performance variation and dangerous inertia when genuine problems emerge.

Adaptation decisions should follow a disciplined process that begins with root cause analysis. When monitoring reveals performance issues, organizations must determine whether problems stem from the AI model itself, the data feeding the system, the human oversight mechanisms, or the integration architecture connecting these components. A credit approval system denying applications at unexpected rates might reflect biased training data, changing economic conditions that the model hasn't encountered, human reviewers applying inconsistent override criteria, or technical integration failures preventing the system from accessing updated information.

The adaptation process must preserve institutional memory by documenting what was changed, why adjustments were made, and what results occurred.[1] This creates organizational learning that prevents repeatedly encountering the same problems and enables knowledge transfer as personnel change. Organizations that treat each adaptation as an isolated technical fix rather than a learning opportunity fail to develop the institutional wisdom required for responsible long-term AI stewardship.

Critically, continuous monitoring and adaptation require protecting human judgment from the efficiency pressures that AI automation creates.[34] Organizations must resist the temptation to automate the monitoring and adaptation processes themselves, recognizing that human oversight of AI systems cannot be delegated to other AI systems without creating recursive accountability problems. The goal is to establish a sustainable rhythm of observation, reflection, and adjustment that keep AI systems aligned with organizational values and stakeholder needs as both technology and context evolve. The pharmaceutical company that opened this chapter represents a cautionary tale that organizations across industries continue to repeat: investing heavily in AI technology while neglecting the human, organizational, and ethical infrastructure required for successful adoption. The transformation roadmap presented here offers an alternative path—one that recognizes AI implementation as fundamentally an organizational challenge rather than merely a technical deployment.

The framework's progression from readiness assessment through governance establishment, pilot testing, scaling with safeguards, capability building, and continuous adaptation reflects a critical insight that responsible AI adoption cannot be rushed. Organizations that skip foundational steps like assessing cultural readiness, establishing accountability structures and testing ethical implications at small scale inevitably encounter the problems this deliberate approach prevents. The expensive algorithms sitting underutilized, the accountability vacuums when systems cause harm, the employee resistance that undermines technical capabilities—these failures stem not from inadequate technology but from inadequate preparation.

The governance-first principle stands as perhaps the most counterintuitive yet essential element of the roadmap. Conventional technology adoption follows a familiar sequence: identify business need, select technology, deploy system, then address governance gaps as they emerge. AI decision-making systems demand inverting this sequence because the stakes of getting governance wrong such as discriminatory outcomes, privacy violations, and accountability vacuums manifest too quickly and cause too much harm to be addressed reactively. Establishing clear lines of responsibility, defining decision rights between humans and machines, and translating ethical principles into operational policies before deployment transforms governance from compliance burden into design constraint that shapes what gets built and how it operates.

The pilot program strategy similarly challenges conventional implementation wisdom by insisting that technical performance metrics alone provide insufficient evidence for scaling decisions. Organizations must evaluate whether systems operate fairly across demographic groups, whether explanations enable meaningful human oversight, and whether affected stakeholders experience the implementation as procedurally just. These ethical dimensions cannot be retrofitted after enterprise-wide deployment. They must be tested, measured, and validated at pilot scale when course correction remains feasible.

As organizations move from controlled pilots to enterprise deployment, the framework addresses the fundamental tension between operational efficiency and meaningful human oversight. The differentiated governance models—human-in-the-loop for high-stakes decisions, human-over-the-loop for moderate-risk contexts, and hybrid approaches with escalation protocols—recognize that not all decisions warrant identical oversight intensity. This nuanced approach enables scaling while preserving human agency where it matters most, working with rather than against human cognitive limitations.

The emphasis on building stewardship capabilities and continuous monitoring reflects the reality that AI systems are not static deployments but dynamic capabilities requiring ongoing management. Organizations must develop institutional competencies for recognizing when systems drift from intended behavior, when organizational contexts shift in ways that render initial design decisions obsolete, and when adaptation becomes necessary. This transforms AI governance from a project phase with a defined endpoint into an operational discipline embedded in organizational functioning.

For executives, consultants, and policymakers navigating the practical realities of AI adoption, this roadmap offers more than a checklist of implementation steps. It provides a framework for thinking differently about what AI transformation requires—not just technical capabilities but cultural readiness, not just algorithmic sophistication but governance maturity, not just deployment speed but sustained stewardship. The organizations that succeed in adopting AI decision-making systems responsibly will be those that recognize the technology as catalyst for broader organizational transformation rather than a standalone technical project. The roadmap presented here provides the structure for making that transformation deliberate, ethical, and sustainable.

CHAPTER 7.

DESIGNING MORAL MACHINES: CREATING SHARED FRAMEWORKS FOR A HUMAN-CENTERED FUTURE

As autonomous vehicle manufacturers faced an impossible engineering challenge to program the decision-making system to handle unavoidable accident scenarios where someone would be harmed regardless of the choice made. The engineers could write the code, but they couldn't answer the underlying question as to whose values should the machine embody when the stakes involved life and death, and who had the authority to make that determination for millions of users across different cultures, legal systems, and moral traditions?

This scenario, far from hypothetical, represents the frontier challenge confronting organizations and societies as AI systems transition from operational tools to entities that must navigate moral complexity. The previous chapters have explored how AI decision-making systems function, the accountability gaps they create, and the governance frameworks organizations can implement. Yet these practical considerations ultimately converge on a more fundamental question: how do we design AI systems that reflect shared human values when those values themselves remain contested, contextual, and culturally diverse?

The challenge extends well beyond autonomous vehicles. Financial algorithms must balance profit maximization against community investment and economic stability. Healthcare AI systems navigate trade-offs between efficiency, equity, and individual patient circumstances. Criminal justice algorithms weigh public safety against rehabilitation and proportionality. In each domain, the technical capacity to automate decisions has outpaced our collective agreement about what principles should guide those choices.

This chapter examines how organizations, policymakers, technologists, and ethicists can collaborate to create frameworks for AI decision-making that transcend individual corporate interests while remaining grounded in practical implementation realities. The goal is not to achieve universal consensus on every ethical question, which is an impossible standard even for human decision-makers, but rather to establish processes, governance structures, and accountability mechanisms that ensure AI systems remain responsive to human values as those values evolve and vary across contexts.

The path forward requires moving beyond two inadequate approaches that have dominated early AI ethics discussions. The first treats ethics as a purely technical problem, assuming that moral principles can be reduced to mathematical optimization functions that engineers simply code into algorithms. The second treats ethics as entirely separate from technical design, relegating moral considerations to policy documents and oversight committees that remain disconnected from actual system development. Neither approach adequately addresses the reality that designing moral machines requires sustained collaboration between those who understand technical capabilities and those who can articulate ethical implications.

Several promising models have emerged from organizations and jurisdictions attempting to bridge this gap. Multi-stakeholder governance frameworks bring together diverse perspectives in system design phases rather than after deployment. Participatory design processes engage communities who will be affected by AI decisions in defining acceptable system behavior. International standards initiatives seek common principles that can guide development across cultural contexts while allowing for contextual adaptation. Adaptive governance structures create mechanisms for ethical frameworks to evolve as both technology and societal values change.

These approaches share a common recognition. The future of AI decision-making depends less on the sophistication of the algorithms themselves than on the quality of human systems—governance structures, accountability mechanisms, participatory processes, and institutional safeguards—we build around them. The most technically advanced AI system will fail ethically if deployed within inadequate governance frameworks, while even relatively simple automated decision-making can operate responsibly when embedded in robust accountability structures.

This chapter synthesizes insights from the preceding discussions into a forward-looking vision for human-centered AI governance. It examines how we can preserve human agency, protect dignity, and promote collective wellbeing while capturing AI's transformative potential—not through perfect solutions, but through ongoing commitment to inclusive dialogue, adaptive frameworks, and the recognition that designing moral machines ultimately requires us to clarify and strengthen our own moral architectures.

The Value Alignment Challenge: Encoding Ethics When Human Values Diverge Across Cultures and Contexts

The technical challenge of value alignment (ensuring AI systems behave consistently with human values) becomes exponentially more complex when those values themselves diverge across cultures, contexts, and communities. A healthcare AI system trained on Western medical ethics emphasizing patient autonomy may operate inappropriately in contexts where family-centered decision-making predominates. A content moderation algorithm reflecting one society's speech norms may systematically suppress legitimate expression in another. These are not hypothetical edge cases but fundamental design challenges confronting every organization deploying AI systems across diverse populations.

The World Economic Forum's research on AI value alignment emphasizes that while certain principles like privacy enjoy widespread recognition, their practical interpretation differs significantly between regions.[89] Some jurisdictions prioritize individual data rights, while others emphasize collective security or community welfare.[89 90] This divergence means that a uniform global approach to encoding values into AI systems may be neither realistic nor desirable[89 90], yet purely localized approaches risk creating fragmented systems that cannot operate across borders or serve diverse populations within single jurisdictions.

Current technical approaches to value alignment struggle with this pluralism. Crowdsourcing methods that aggregate human judgments to shape AI behavior can approximate widely shared norms within specific populations, but they systematically disadvantage minority perspectives through majoritarian aggregation.[91] Reinforcement learning from human feedback, which is the technique behind many advanced language models, trains systems to match the preferences of particular rater groups, effectively encoding those specific values as if they were universal.[91] Constitutional AI, which uses explicit written principles to guide system behavior, makes value choices more transparent but cannot resolve the underlying question of which principles should govern when reasonable people disagree.[91]

Recent philosophical analysis highlights that these methods fail to accommodate reasonable moral disagreement—the persistent ethical differences that remain even among informed, reflective, and well-intentioned people.[91] The techniques produce morally controversial outputs without providing either strong epistemic reasons to believe those outputs are correct or legitimate political reasons for diverse populations to accept them.[91] This limitation reveals that value alignment is not purely a technical problem but fundamentally involves questions of moral epistemology and political legitimacy.[91 92]

Despite these challenges, several normative anchors offer partial solutions. International human rights frameworks provide a baseline floor of protections that apply across cultural contexts, even as their interpretation varies.[90] The concept of ethical "red lines"—non-negotiable boundaries that AI systems must never cross—helps define outer limits while allowing legitimate diversity within those constraints.[89 90] Red lines might include prohibitions on impersonating humans without disclosure, unauthorized self-modification that violates safety constraints, or actions that directly violate fundamental rights.[89 90] Within these boundaries, context-sensitive design becomes essential.

The World Economic Forum advocates tailored approaches that adapt AI systems to specific cultural, legal, and societal contexts rather than imposing one-size-fits-all solutions.[89 90] A credit-scoring algorithm deployed across diverse regions requires training on localized datasets reflecting regional financial behaviors and regular auditing with fairness metrics to detect discrimination against particular groups.[89] A healthcare diagnostic system must balance patient autonomy, collective welfare considerations, privacy expectations, and regulatory requirements differently depending on deployment context, while maintaining sufficient transparency for doctors and patients to understand and appropriately trust its recommendations.[89 90]

This contextualization requires robust governance structures. Multi-stakeholder engagement processes that include affected communities—

particularly historically marginalized groups—help ensure alignment reflects diverse perspectives rather than narrow elite interests.[89] [90] Organizational AI management systems guided by standards like ISO/IEC 42001 embed ethical considerations throughout development lifecycles.[89] [90] Regular technical and ethical audits assess whether systems maintain alignment as both technology and societal values evolve.[89 90]

The value alignment challenge ultimately demands recognition that designing moral machines requires ongoing dialogue about human values themselves, institutional mechanisms for legitimate decision-making about contested norms, and adaptive frameworks that can evolve rather than locking in particular moral perspectives permanently.[92]

Multi-Stakeholder Governance Models: Building Inclusive Frameworks That Represent Diverse Perspectives

The complexity of AI decision-making systems demands governance structures that extend beyond traditional corporate hierarchies or regulatory bodies. Multi-stakeholder governance models have emerged as a practical response to this challenge, bringing together diverse perspectives from technical experts, business leaders, ethicists, legal professionals, affected communities, and civil society organizations into collaborative frameworks that shape how AI systems are designed, deployed, and monitored.[93] These models recognize that no single group possesses the complete expertise or legitimacy to govern AI systems that affect multiple constituencies across different contexts.

The Partnership on AI, established in 2016 by major technology companies and later expanded to include nonprofits, academic institutions, and civil society organizations, exemplifies this approach. The partnership convenes stakeholders across sectors to develop best practices, conduct research, and provide guidance on AI development that reflects diverse perspectives rather than narrow corporate interests. This model demonstrates that multi-stakeholder governance can function at scale, producing frameworks like their guidance on algorithmic impact assessments that organizations worldwide have adopted.

Multi-stakeholder models address a fundamental legitimacy challenge that single-entity governance cannot resolve. When an AI system makes decisions affecting healthcare access, employment opportunities, or criminal justice outcomes, those affected by these decisions have legitimate claims to participate in defining acceptable system behavior. The IEEE's Global Initiative on Ethics of Autonomous and Intelligent Systems brings together engineers, ethicists, policymakers, and community representatives to develop technical standards that embed ethical considerations into system design—recognizing that technical standards are not value-neutral but embody choices about whose interests matter and how competing values should be balanced.

Effective multi-stakeholder governance requires more than simply assembling diverse participants. The structure must provide meaningful influence rather than token representation. Research from the Ada Lovelace Institute on public participation in AI governance emphasizes that inclusive frameworks must address power imbalances between well-resourced technology companies and under-resourced civil society organizations or community groups. This requires providing resources for participation, ensuring technical information is accessible to non-experts, and creating decision-making processes where all stakeholders can genuinely influence outcomes rather than merely commenting on predetermined conclusions.

The European Union's approach to AI regulation illustrates multi-stakeholder governance at the policy level. The development of the AI

Act involved extensive consultation with technology companies, civil society organizations, academic researchers, and member state governments—each bringing different priorities and concerns. Industry representatives emphasized innovation and competitiveness, civil rights organizations focused on fundamental rights protection, and governments balanced economic interests against citizen welfare. The resulting framework reflects compromises and trade-offs negotiated through this multi-stakeholder process, producing regulation with greater legitimacy and practical applicability than any single constituency could have designed alone.

Corporate governance structures increasingly incorporate multi-stakeholder models through AI ethics boards and review committees that include external members alongside company representatives.[16] Microsoft's AI Ethics Committee, for example, includes external advisors with expertise in ethics, law, and social science who review high-stakes AI applications before deployment. These structures create accountability mechanisms where decisions must be justified to stakeholders beyond those with direct financial interests in rapid deployment.[94]

The challenge for multi-stakeholder governance lies in maintaining effectiveness while ensuring inclusivity. Consensus-building across diverse perspectives takes time, potentially slowing AI development in competitive markets.[93] Organizations must balance thoroughness with practicality, creating governance processes that genuinely incorporate diverse perspectives without becoming paralyzed by the need for universal agreement.[17] This requires establishing clear decision-making protocols, defining which decisions require full stakeholder consensus versus consultation, and building trust through transparent processes where stakeholder input demonstrably influences outcomes rather than serving as performative inclusion.[94]

From Corporate Standards to Global Principles: Emerging International Approaches to AI Ethics and Regulation

The limitations of multi-stakeholder governance become particularly apparent when AI systems operate across national boundaries, where different legal traditions, cultural values, and regulatory philosophies collide. A content moderation algorithm trained to reflect one society's speech norms may systematically suppress legitimate expression in another. A credit-scoring system designed under one jurisdiction's fairness standards may violate anti-discrimination laws elsewhere. These conflicts have catalyzed a shift from localized corporate standards toward coordinated international frameworks that seek common principles without imposing uniform solutions.[84 85]

The European Union has pioneered the most comprehensive approach to translating ethical principles into binding regulation. The EU's AI Act, developed through extensive consultation with technology companies, civil society organizations, academic researchers, and member state governments, establishes a risk-based regulatory framework that scales obligations according to the potential harm AI systems pose to fundamental rights, health, and safety.[84] High-risk applications—including biometric identification, credit scoring, employment decisions, and critical infrastructure control—face strict requirements for data governance, transparency, human oversight, and robustness.[84] Certain uses are prohibited entirely, such as systems that manipulate vulnerabilities or undermine democratic participation.[84]

This European approach reflects a broader philosophical commitment to grounding AI governance in human rights frameworks rather than purely economic or innovation-focused considerations.[84] By treating AI regulation as an extension of existing fundamental rights protections, in particular, privacy, equality, due process, and freedom of expression, the EU creates normative continuity between constitutional traditions and emerging technological challenges.[84] The emphasis extends beyond preventing individual harms to preserving democratic institutions and social cohesion against large-scale, opaque algorithmic systems.[84]

The United States has pursued a markedly different path, favoring sectoral and decentralized governance over comprehensive AI-specific legislation.[84] Different regulatory agencies issue AI-related guidance within their existing mandates. Financial regulators address algorithmic trading and credit decisions, health agencies oversee medical AI applications, employment regulators examine hiring algorithms—creating a fragmented landscape where obligations vary significantly by domain.[84] This approach prioritizes maintaining competitive advantage and innovation leadership while addressing high-profile harms through existing legal frameworks for consumer protection, anti-discrimination, and sector-specific safety requirements.

Beyond these major regulatory powers, international organizations have attempted to identify shared principles that can inform national laws and corporate practices across diverse contexts. These efforts increasingly ground AI governance in universal human rights, treating AI's impact on privacy, equality, and political participation as extensions of existing international treaty obligations rather than entirely novel challenges.[84] This human rights framing provides cross-cultural legitimacy and creates bridges between national constitutions, international law, and AI-specific policies.[84]

The challenge lies in translating broad principles into operational, testable, and auditable requirements.[85] Technical standard-setting bodies are developing norms for AI lifecycle processes—from data collection and labeling to deployment and monitoring—that make ethical aims actionable. Fairness metrics quantify disparate impact across demographic groups.[85] Explainability standards require documentation that allows stakeholders to understand system behavior.[85] Privacy-by-design practices embed data minimization and access controls into system architecture.[85] These technical standards represent attempts to bridge the gap between abstract ethical commitments and concrete engineering practices.

Yet significant tensions persist. Philosopher Mark Coeckelbergh emphasizes that intellectual power over AI "is concentrated within a few big corporations that will need to be regulated in order to safeguard the public interest," highlighting the challenge of democratic oversight when technical expertise and infrastructure remain concentrated in private hands.[84] The balance between innovation incentives and rights protection varies dramatically across jurisdictions, creating opportunities for regulatory arbitrage where companies migrate operations to more permissive environments. Cross-border data flows and computational resources mean that AI systems increasingly operate in a jurisdictional gray zone where no single regulatory framework fully applies.[84]

The emerging international landscape thus reflects ongoing negotiation between competing priorities: innovation versus regulation, corporate autonomy versus democratic accountability, universal principles versus cultural pluralism.[84 85] The trajectory points toward greater coordination and harder law, but implementation remains uneven and contested.[84 85]

Participatory Design for Moral Machines: Engaging Communities in Defining Acceptable AI Behavior

The theoretical frameworks for AI ethics gain practical traction only when those affected by algorithmic decisions participate directly in defining what constitutes acceptable system behavior. Participatory design, the intentional engagement of diverse stakeholders in shaping technology development—offers a methodological bridge between abstract ethical principles and operational AI systems that communities perceive as legitimate and trustworthy. This approach recognizes that moral machines cannot be designed in isolation by technical teams or ethics committees alone but require sustained collaboration with the people whose lives these systems will shape.

Participatory design for AI systems moves beyond traditional consultation models where stakeholders provide feedback on predetermined options. Research from the Ada Lovelace Institute distinguishes between shallow engagement—informing or consulting stakeholders—and deeper forms that involve collaboration or empowerment, where affected communities co-create design requirements and share decision-making authority. For moral machines, this deeper engagement proves essential because technical teams, however well-intentioned, cannot fully anticipate how algorithmic decisions will affect diverse populations across different contexts or identify the subtle ways systems might reproduce existing inequities.

The practical application of participatory design in AI development takes several forms. Co-design workshops bring together technologists, domain experts, and community members to collectively define problems, map values, and establish design constraints before system development begins. These sessions surface implicit assumptions that technical teams might otherwise encode into algorithms without recognizing their contested nature. A healthcare AI system designed without patient participation might optimize for efficiency metrics that conflict with patient preferences for autonomy or family involvement in treatment decisions—conflicts that become visible only when diverse stakeholders articulate competing values explicitly.

Recent implementations demonstrate participatory design's capacity to ground ethical principles in operational requirements. A public health initiative deploying AI tools for diabetes management established partnership tables including policymakers, healthcare providers, and community representatives who conducted ethical audits of AI models to identify potential biases, co-designed data pipelines and user interfaces, and defined success metrics emphasizing trustworthiness and acceptability rather than purely technical performance measures.[95] This sustained engagement created governance structures where community input shapes ongoing system adaptation rather than serving as one-time consultation.[95]

The Virtual World Café method exemplifies scalable participatory approaches that accommodate large, diverse stakeholder groups.[96] Small teams rotate through discussion tables focused on different AI design dimensions—fairness definitions, transparency requirements, accountability mechanisms—allowing participants to contribute expertise while learning from others' perspectives.[96] This structured dialogue surfaces the contextual nature of ethical principles: what fairness means in employment algorithms differs from its meaning in credit scoring, and only those affected can fully articulate these distinctions.

Design fictions—plausible scenarios depicting future AI applications—provide another participatory tool for exploring ethical boundaries.[97] When communities respond to speculative prototypes of AI systems in criminal justice or social services, they reveal implicit norms about acceptable automation, identify decisions that should never be fully delegated to machines, and propose governance mechanisms like oversight boards or appeal processes.[97] These exercises translate abstract principles into concrete behavioral constraints that developers can implement.

Yet participatory design faces significant implementation challenges. Power imbalances persist even in collaborative settings, where technical expertise and institutional authority can override community input without careful facilitation. The method requires substantial time and resources that resource-constrained organizations struggle to provide. AI systems' technical complexity creates barriers to meaningful participation unless developers invest in translation, simplification, and co-learning that bridges the gap between technical and civic knowledge. The temporal mismatch between rapid AI development cycles and slower participatory processes demands new approaches that align iterative system evolution with ongoing community engagement.[96]

Despite these challenges, participatory design remains essential for creating AI systems that embody shared values rather than narrow technical or corporate interests.[98] The legitimacy of moral machines ultimately depends not on algorithmic sophistication but on whether affected communities recognize these systems as accountable to their values and responsive to their concerns.[95]

Adaptive Ethics Frameworks: Creating Governance Systems That Evolve with Technology and Societal Values

The most sophisticated ethical framework becomes obsolete the moment technology or societal expectations shift beyond its original design parameters. A governance structure that adequately addressed facial recognition concerns in 2018 may prove inadequate for generative AI systems in 2024, just as today's frameworks will struggle with technologies not yet conceived.[99 101 103] This temporal challenge—the mismatch between static governance and dynamic reality—has catalyzed development of adaptive ethics frameworks that treat AI governance not as a fixed rulebook but as an evolving system capable of learning and adjusting alongside the technologies it oversees.[99 100 103]

Adaptive ethics frameworks represent a fundamental reconceptualization of how organizations and societies govern AI decision-making systems. Rather than establishing comprehensive rules intended to remain stable indefinitely, these frameworks create institutional mechanisms for continuous monitoring, assessment, and revision of ethical standards in response to technological change, emerging harms, and shifting societal values.[99 100 103] UNESCO's Recommendation on the Ethics of Artificial Intelligence explicitly calls for "multi-stakeholder and adaptive governance and collaboration" as a core principle, recognizing that effective oversight requires structures designed for evolution rather than permanence.[104]

The European Union's AI Act demonstrates this adaptive approach at the regulatory level. The legislation establishes a risk-based framework that categorizes AI systems by potential harm—unacceptable risk, high risk, limited risk, and minimal risk—with requirements scaled accordingly.[102 35] Critically, the Act's structure allows regulatory authorities to update the list of high-risk applications as new technologies and use cases emerge, avoiding the obsolescence that plagues static regulation.[99] This design forces organizations to build internal governance processes capable of ongoing adjustment rather than one-time compliance, effectively institutionalizing adaptivity throughout the AI ecosystem.[35 103]

Organizations implementing adaptive frameworks typically employ modular governance architectures where different ethical dimensions—privacy, fairness, transparency, safety—function as semi-independent modules that can be updated without overhauling the entire system.[100] A financial institution might revise its bias mitigation module to comply with new fairness standards in credit scoring while keeping privacy and security modules stable.[100] This modularity enables targeted adaptation in response to specific regulatory changes, technical advances, or identified harms without creating systemic disruption.[100]

The operational core of adaptive ethics frameworks consists of continuous monitoring and feedback mechanisms that detect when existing governance proves inadequate.[99 100 103] Organizations establish interdisciplinary AI governance committees empowered to review high-risk projects, respond to incidents, and periodically update policies and technical standards.[100 101 103] These committees typically include legal, technical, ethical, and business operations expertise, reflecting recognition that no single discipline possesses complete perspective on AI's implications.[100 101] Fractal Analytics describes this approach as cognitive compliance—embedding regulatory, ethical, and organizational standards directly into AI systems with modular components that can be updated as rules evolve without retraining entire models.[100]

Yet adaptive frameworks face inherent tensions. Research on AI governance warns that excessive reliance on self-regulation and voluntary adaptation can create gaps in protection, particularly where commercial incentives conflict with public interest.[101 103] Frequent but superficial rule changes may constitute "ethics theater" that fails to address structural harms.[101] The challenge lies in distinguishing genuine adaptation—substantive revision in response to evidence of inadequacy—from performative adjustment that maintains appearances while avoiding meaningful change.

Effective adaptive frameworks therefore require anchoring in non-negotiable baseline protections that remain stable even as operational details evolve.[101 104] International human rights frameworks, constitutional protections, and fundamental legal principles provide this stability, establishing outer boundaries within which adaptation occurs. The framework adapts not by weakening core commitments to dignity, equality, and due process, but by refining how these principles apply to novel technological contexts and by updating technical mechanisms for achieving them as capabilities advance.[101 104] This combination of stable values with evolving implementation enables governance systems to remain both principled and responsive in environments where neither technology nor society stands still.[99 101 104]

The Human-Centered Future: Preserving Agency, Dignity, and Collective Wellbeing in an AI-Augmented World

The ultimate measure of success in AI governance lies not in the sophistication of algorithms or the elegance of ethical frameworks, but in whether these systems preserve and enhance the fundamental conditions for human flourishing: agency, dignity, and collective wellbeing. As AI decision-making systems become increasingly capable and pervasive, the challenge shifts from whether to adopt these technologies to how we can ensure they serve genuinely human-centered purposes rather than optimizing narrow objectives that undermine broader values.[107 109]

Stuart Russell's critique of what he terms the "standard model" of AI development illuminates this challenge with particular clarity.[107] The prevailing approach treats machines as intelligent to the extent that their actions achieve their programmed objectives, creating systems that optimize fixed goals with increasing effectiveness.[107] Yet this paradigm systematically under-specifies human values, producing algorithms that accomplish their technical objectives while generating outcomes humans never endorsed.[107 108 109] Social media recommendation systems maximize engagement by amplifying divisive content.[109] Predictive policing tools achieve statistical accuracy while reinforcing patterns of discriminatory enforcement. Credit-scoring algorithms optimize default prediction while systematically disadvantaging communities whose historical data reflects structural inequities rather than creditworthiness.

These failures share a common architecture: systems designed to pursue predetermined objectives without uncertainty about whether those objectives genuinely reflect human preferences or societal values.[107 109]

Russell proposes a fundamental reconceptualization where machines are designed to maximize realization of human preferences while remaining uncertain about what those preferences actually are.[105 110] This uncertainty becomes a design feature rather than a defect, creating systems that defer to human judgment, seek clarification when facing ambiguous situations, and treat human intervention—including being shut down—as informative signals about true objectives rather than obstacles to overcome.[105 107 110]

The practical implications of this shift extend beyond technical architecture to questions of control and autonomy. In Russell's formal models, including the Off-Switch Game, AI systems that recognize their uncertainty about human preferences willingly allow shutdown because they understand that human intervention likely indicates misalignment between current actions and actual human goals.[105] This contrasts sharply with standard fixed-objective systems, which would instrumentally resist being turned off since shutdown prevents them from maximizing their programmed utility.[106 108] The difference illustrates how moral architectures—the combination of technical design, institutional governance, and cultural values—shape whether AI systems preserve or erode human agency.

Preserving dignity in AI-augmented environments requires moving beyond treating individuals as data points in optimization functions. When people become features in datasets or labels in classification systems, decisions can ignore the uniqueness of persons and their circumstances. Algorithmic discrimination in hiring, lending, and criminal justice violates the equal moral status that dignity demands. Pervasive surveillance and predictive analytics create environments of continuous monitoring that chill autonomy and self-expression. A human-centered approach embeds non-discrimination constraints into system design, ensures transparency appropriate to context, protects privacy by design, and creates avenues for voice and redress when individuals experience algorithmic harm.[105 107]

Collective wellbeing demands attention to systemic effects that extend beyond individual interactions with AI systems. Concentration of economic power through automation and data network effects can exacerbate inequality. Micro-targeted political messaging administered by AI can fragment public discourse and undermine democratic processes. Large-scale labor displacement without adequate policy response destabilizes communities and reduces meaningful work opportunities. Russell frames the long-term challenge through what he calls the "gorilla problem"—just as gorillas have essentially no future beyond what humans allow, humanity faces questions about whether it can maintain autonomy in a world with machines of substantially greater intelligence if those systems optimize misaligned objectives.[105 106 110]

The path toward a genuinely human-centered future requires moral architectures that span technical algorithms, institutional frameworks, and cultural narratives about what AI should serve. These architectures must integrate individual preferences with shared public values, rights-based constraints that protect minorities and fundamental freedoms, and procedural fairness in how decisions are made. The challenge is not merely technical but fundamentally concerns what kind of world humans want to create and whether the systems being built today will preserve or foreclose the agency, dignity, and collective flourishing that define a life worth living. The autonomous vehicle scenario that opened this chapter with engineers capable of writing code but unable to determine whose values that code should embody captures the essential challenge confronting organizations and societies as AI systems transition from tools to decision-makers navigating moral complexity. The preceding sections have examined how value alignment becomes exponentially more difficult when human values themselves diverge across cultures and contexts, how multi-stakeholder governance models attempt to build inclusive frameworks that represent diverse perspectives, and how international approaches seek common principles without imposing uniform solutions. These explorations converge on a fundamental recognition: designing moral machines ultimately requires clarifying and strengthening our own moral architectures—the human systems of governance, accountability, and shared purpose we build around the algorithms.

The path forward demands moving beyond two inadequate extremes that have dominated early AI ethics discussions. Treating ethics as a purely technical problem by assuming moral principles can be reduced to optimization functions that engineers simply code ignores the reality that reasonable people disagree about fundamental values even when informed and reflective. Conversely, treating ethics as entirely separate from technical design and relegating moral considerations to policy documents disconnected from actual development ensures that implicit value choices embedded in system architecture remain unexamined and unaccountable. Neither approach adequately addresses the reality that designing moral machines requires sustained collaboration between those who understand technical capabilities and those who can articulate ethical implications.

The frameworks examined throughout this chapter share several essential characteristics that distinguish genuinely human-centered approaches from performative ethics. They establish participatory processes that engage affected communities in defining acceptable AI behavior rather than imposing predetermined solutions. They create adaptive governance structures capable of evolving as both technology and societal values change, avoiding the obsolescence that plagues static regulation. They anchor flexibility in non-negotiable baseline protection fundamental rights, human dignity, democratic participation—that remain stable even as operational details adapt to new contexts. They recognize that legitimacy depends not on algorithmic sophistication but on whether diverse stakeholders perceive systems as accountable to their values and responsive to their concerns.

The technical dimensions of this challenge remain significant. Value alignment methodologies must accommodate reasonable moral disagreement rather than aggregating preferences in ways that systematically disadvantage minority perspectives. Transparency mechanisms must balance explainability requirements with competitive concerns and privacy protections. Monitoring systems must detect when AI behavior diverges from intended values before causing substantial harm. Yet these technical challenges exist within broader institutional contexts that ultimately determine whether AI systems preserve or erode human agency, dignity, and collective wellbeing.

The future of AI decision-making depends less on the sophistication of the algorithms themselves than on the quality of human systems—governance structures, accountability mechanisms, participatory processes, and institutional safeguards—we build around them. The most technically advanced AI system will fail ethically if deployed within inadequate governance frameworks, while even relatively simple automated decision-making can operate responsibly when embedded in robust accountability structures that preserve meaningful human oversight and intervention.

This chapter has argued that designing moral machines requires ongoing social dialogue, inclusive participation in framework development, and institutional mechanisms that allow ethical standards to evolve alongside technology and societal values. The challenge is not to achieve universal consensus on every ethical question—an impossible standard even for human decision-makers—but rather to establish processes that ensure AI systems remain responsive to human values as those values evolve and vary across contexts. The autonomous vehicle manufacturer's engineers could write the code, but the deeper question of whose values should guide unavoidable trade-offs demands answers that only inclusive, adaptive, and genuinely human-centered governance can provide.

CONCLUSION

The autonomous vehicle manufacturer's dilemma of how to program moral choices into machines represents more than an engineering challenge. It crystallizes the fundamental question this book has explored: not whether AI will make decisions that affect human lives, but how we will govern those systems to ensure they remain accountable to human values, preserve meaningful agency, and serve collective wellbeing rather than narrow optimization metrics.

Throughout these chapters, we've examined AI decision-making systems as they currently operate in boardrooms, healthcare facilities, financial institutions, and supply chains. The evidence reveals a consistent pattern: organizations that treat AI adoption purely as a technical implementation inevitably encounter accountability gaps, ethical blind spots, and erosion of human agency. Conversely, those that approach AI as a governance challenge—establishing clear responsibility structures, maintaining meaningful human oversight, and building adaptive ethical frameworks—capture competitive advantages while preserving the human judgment that algorithms cannot replicate.

The insurance company that couldn't explain its claim denials, the retailer whose algorithm detected patterns invisible to experienced analysts, the healthcare system whose AI recommendations went unquestioned until patients were harmed—these examples demonstrate that maintaining human accountability when machines make decisions requires deliberate architectural choices. It demands governance structures established before deployment, transparency mechanisms that enable meaningful oversight, and organizational cultures that empower humans to question and override algorithmic recommendations when contextual judgment demands it.

For corporate strategists and executives navigating digital transformation, the path forward requires balancing AI's pattern recognition capabilities with human contextual understanding. The competitive pressure to automate decision-making is real, yet the organizations that will thrive long-term are those that resist the temptation to optimize purely for speed and efficiency. Instead, they design systems that preserve intervention points where human judgment can supersede algorithmic recommendations, establish clear accountability for AI-influenced decisions, and cultivate leadership competencies that integrate algorithmic literacy with traditional executive virtues like ethical discernment and stakeholder empathy.

For policymakers and government officials, the challenge lies in crafting regulatory frameworks that protect human dignity and agency without stifling innovation. The emerging international standards and multi-stakeholder governance models examined in this book demonstrate that effective AI regulation requires collaboration across disciplines and borders. It demands participatory approaches that engage diverse communities in defining acceptable AI behavior, adaptive frameworks that evolve as technology and societal values change, and enforcement mechanisms that assign meaningful accountability when systems cause harm.

For professionals across industries concerned with the future of work, the transformation underway is not about humans versus machines but about redefining human roles in AI-augmented environments. The skills that become increasingly valuable are precisely those that algorithms cannot replicate: ethical judgment in ambiguous situations, contextual interpretation of data-driven insights, creative problem-solving that transcends pattern recognition, and the capacity to recognize when human intuition should override statistical recommendations.

The future of AI decision-making depends less on algorithmic sophistication than on the human systems of governance, accountability, and shared purpose we build around these technologies. Designing moral machines requires ongoing social dialogue about whose values these systems embody, inclusive participation in framework development, and institutional mechanisms that ensure AI serves human flourishing rather than narrow efficiency metrics.

The question is no longer whether AI will influence critical decisions affecting millions of lives already does. The question is whether we will develop the governance frameworks, organizational capabilities, and collaborative structures necessary to ensure these systems remain accountable to human values, preserve meaningful agency, and contribute to a future where technology augments rather than diminishes what makes us human. That future is not predetermined by technological capability but shaped by the choices we make today about how to integrate AI into the moral architectures of our organizations and societies.

FEEDBACK

YOUR VOICE TRULY MATTERS

If this book helped you learn something new, feel more confident about your health, or sparked your curiosity, your feedback would mean a great deal.

Please consider sharing your review on Amazon to help others discover this journey.

You can also visit and follow **Alexandra Isaacs on Amazon** to stay connected and explore more insights.

Thank you sincerely for your time, your support, and for being part of this exploration into artificial intelligence.

With gratitude,
Alexandra Isaacs

CONTINUE THE JOURNEY

If you found this book insightful, explore the previous titles of

The Future Series:

• **Book 1**: Beyond the Algorithm

• **Book 2**: The Humanoid Age

Available on Amazon

REFERENCES

1 O'Callaghan, M. (2023, April 26). Decision Intelligence: Human–Machine Integration for Decision-Making. Routledge. Https://www.routledge.com/Decision-Intelligence-Human-Machine-Integration-for-Decision-Making/ocallaghan/p/book/9781032384092

2 Professor. (2025, August 22). How AI Is Transforming Executive Leadership in 2025. Knowledge Blog. Https://thecasehq.com/how-ai-is-transforming-executive-leadership-in-2025/

3 Fourrage, L. (2025, August 2). Top 10 AI Strategies Every Executive Should Know in 2025. Nucamp. Https://www.nucamp.co/blog/ai-essentials-for-work-2025-top-10-ai-strategies-every-executive-should-know-in-2025

4 The Recruitability. (n.d.). The AI Revolution: Business Strategies for Executives in 2025. The Recruitability. Https://www.therecruitability.com/as-artificial-intelligence-ai-evolves-its-transformative-potential-across-industries-grows-exponentially-for-executives-2025-presents-an-unprecedented-opportunity-to-harness-ai-to-drive-innovatio/

5 Accountability Now. (2025, November 14). 9 Game-Changing AI Ideas for ceos in 2025. Accountability Now. Https://accountabilitynow.net/ai-ideas-for-ceos/

6 Futurum Group. (2025, May 16). C-Suite Executives Dominate AI Decision-Making as Strategy Becomes Priority. Futurum Group. Https://futurumgroup.com/press-release/c-suite-executives-dominate-ai-decision-making-as-strategy-becomes-priority/

7 Singla, A., Sukharevsky, A., Hall, B., Yee, L., & Chui, M. (2025, November 5). The state of AI in 2025: Agents, innovation, and transformation. Mckinsey

& Company. Https://www.mckinsey.com/capabilities/quantumblack/our-insights/the-state-of-ai

8 Ransbotham, S., Kiron, D., Khodabandeh, S., Iyer, S., & Das, A. (2025, November 18). The Emerging Agentic Enterprise: How Leaders Must Navigate a New Age of AI. MIT Sloan Management Review. Https://sloanreview.mit.edu/projects/the-emerging-agentic-enterprise-how-leaders-must-navigate-a-new-age-of-ai/

9 Kober, G. (2025, January 24). AI-First Leadership: Embracing the Future of Work. Harvard Business Publishing. Https://www.harvardbusiness.org/insight/ai-first-leadership-embracing-the-future-of-work/

10 Zeiser, J. (2024, June 18). Owning Decisions: AI Decision-Support and the Attributability-Gap. Pubmed Central. Https://pmc.ncbi.nlm.nih.gov/articles/PMC11189344/

11 Lima, G., Grgić-Hlača, N., Jeong, J. K., & Cha, M.. (2022, June 21). The Conflict Between Explainable and Accountable Decision-Making Algorithms. Arxiv. Https://arxiv.org/pdf/2205.05306

12 Munch, L., Mainz, J., & Bjerring, J. C. (2023). The value of responsibility gaps in algorithmic decision-making. Philarchive. Https://philarchive.org/archive/MUNTVO-2

14 Stanford HAI. (n.d.). Designing Decision-Making Algorithms in an Uncertain World. Stanford HAI. Https://hai.stanford.edu/news/designing-decision-making-algorithms-uncertain-world

15 Ujwary-Gil, A., & Florek-Paszkowska, A. (2025). AI, Analytics and Strategic Decision-Making. Routledge. Https://www.routledge.com/AI-Analytics-and-Strategic-Decision-Making/Ujwary-Gil-Florek-Paszkowska/p/book/9781032831107

16 Horseman, M. (2024, December 2). Book of the Month: "AI Governance Comprehensive". DATAVERSITY.

Https://www.dataversity.net/articles/book-of-the-month-ai-governance-comprehensive/

17 Bozdag, E., & Bennati, S. (2025, November). AI Governance: Secure, privacy-preserving, ethical systems. Manning Publications. Https://www.manning.com/books/ai-governance

18 Eitel-Porter, R. (2025). Governing the Machine : How to Navigate the Risks of AI and Unlock Its True Potential. College of Europe. Https://www.coleurope.eu/governing-machine-how-navigate-risks-ai-and-unlock-its-true-potential

19 Eitel-Porter, R., Dongha, P., & Vogel, M. (2025, October 23). Governing the Machine: How to navigate the risks of AI and unlock its true potential. Bloomsbury Publishing. Https://www.bloomsbury.com/us/governing-the-machine-9781399426268/

20 Newsbywire. (October 23, 2025). New book Governing the Machine reveals how to navigate the risks of AI and unlock its true potential. Newsbywire. Https://newsbywire.com/new-book-governing-the-machine-reveals-how-to-navigate-the-risks-of-ai-and-unlock-its-true-potential/

21 Anglen, J. (n.d.). Pattern recognition in ML: A comprehensive overview. Rapid Innovation. Https://www.rapidinnovation.io/post/pattern-recognition-in-ml-a-comprehensive-overview

22 Kirchner, W. B. (2020, February 18). Pattern Recognition and its Role in Decision Making. Exploring the Business Brain. Https://exploringthebusinessbrain.com/pattern-recognition-and-its-role-in-decision-making/

23 huspi. (n.d.). What is pattern recognition in machine learning: guide for business & geeks. Huspi. Https://huspi.com/blog-open/pattern-recognition-in-machine-learning/

24 Baheti, P. (2022, September 13). Pattern Recognition in Machine Learning [Basics & Examples]. V7 Labs. Https://www.v7labs.com/blog/pattern-recognition-guide

25 Operations Council. (2025, January 3). From Reactive to Proactive: How Predictive Analytics is Transforming Operations. Operations Council. Https://operationscouncil.org/from-reactive-to-proactive-how-predictive-analytics-is-transforming-operations/

26 Harrison Clarke. (January 3, 2024). Mastering Decision-Making: Machine Learning for Predictive Analytics. Harrison Clarke. Https://www.harrisonclarke.com/blog/mastering-decision-making-machine-learning-for-predictive-analytics

29 Boros, A. (n.d.). Predictive Modelling. The Decision Lab. Https://thedecisionlab.com/reference-guide/statistics/predictive-modelling

30 Perricos, C., & Kapur, V. (2019, June 24). Anticipatory government: Preempting problems through predictive analytics. Deloitte Insights. Https://www.deloitte.com/us/en/insights/industry/government-public-sector-services/government-trends/2020/predictive-analytics-in-government.html

31 Gesser, A., Gressel, A. R., Xu, M., & Allaman, S. J.. (2022, November 14). When Humans and Machines Disagree – The Myth of "AI Errors" and Unlocking the Promise of AI Through Optimal Decision Making. Debevoise Data Blog. Https://www.debevoisedatablog.com/2022/11/14/when-humans-and-machines-disagree-the-myth-of-ai-errors-and-unlocking-the-promise-of-ai-through-optimal-decision-making-adm-algorithm/

32 O'Callaghan, M. (2023, April 25). Decision Intelligence: Human–Machine Integration for Decision-Making. Taylor & Francis. Https://www.taylorfrancis.com/books/mono/10.1201/b23322/decision-intelligence-miriam-callaghan

33 O'Callaghan, M. (2023, April 26). Decision Intelligence: Human-Machine Integration for Decision-Making. Barnes & Noble.

Https://www.barnesandnoble.com/w/decision-intelligence-miriam-ocallaghan/1142707898

34 Einhorn, C. S. (2025, October 31). When Working With AI, Act Like a Decision-Maker—Not a Tool-User. Harvard Business Review. Https://hbr.org/2025/10/when-working-with-ai-act-like-a-decision-maker-not-a-tool-user

35 AI21 Labs. (n.d.). AI governance frameworks. AI21 Labs. Https://www.ai21.com/knowledge/ai-governance-frameworks/

36 ewsolutions. (n.d.). AI Governance Frameworks: What They Are and How to Use Them. Ewsolutions. Https://www.ewsolutions.com/ai-governance-frameworks/

37 Elevate. (2025, December 2). AI Governance Frameworks Overview: Which Model Is Right?. Elevate Consult. Https://elevateconsult.com/insights/ai-governance-frameworks-overview-which-model-is-right/

38 Pantangi, A. (n.d.). The Architecture of Responsible AI: Balancing Innovation and Accountability. Architecture and Governance. Https://www.architectureandgovernance.com/applications-technology/the-architecture-of-responsible-ai-balancing-innovation-and-accountability/

39 Tunguz, T. (n.d.). The Accountability Machine. Tom Tunguz. Https://tomtunguz.com/the-accountability-machine/

40 H2O Open Casebook. (n.d.). The Ethics and Governance of Artificial Intelligence. H2O Open Casebook. Https://opencasebook.org/casebooks/74426-the-ethics-and-governance-of-artificial-intelligence

41 Wright, M. (2023, November 28). AI Ethics & Governance - A Legal Perspective Vol. 1. Barnes & Noble. Https://www.barnesandnoble.com/w/ai-ethics-governance-a-legal-perspective-vol-1-mark-wright/1144385243

42 Tejano, S. (2025, March 1). Machine Learning Ethics: How to Build AI That's Fair, Transparent, and Accountable. Dart AI. Https://www.dartai.com/blog/machine-learning-ethics

43 trivuma. (october 12, 2025). When machines decide: the quiet urgency of ai ethics. Triumva. Https://triumva.ca/the-power-of-intelligent-thinking-9/

44 Oxford University Press. (n.d.). Chapter Abstract. Oxford Academic. Https://academic.oup.com/edited-volume/34287/chapter/290661457

45 Floridi, L. (2021, November 2). Ethics, Governance, and Policies in Artificial Intelligence. Google Books. Https://books.google.com/books/about/ethicsgovernanceandpoliciesin_Artifi.html?Id=GSFMEAAAQBAJ

46 Raisinghani, M. S. (2025, March 5). 12.3 Ethics of Artificial Intelligence Development and Machine Learning. Openstax. Https://openstax.org/books/foundations-information-systems/pages/12-3-ethics-of-artificial-intelligence-development-and-machine-learning

47 Verhulst, S. (n.d.). Machine Ethics: The Design and Governance of Ethical AI and Autonomous Systems. The Living Library. Https://thelivinglib.org/machine-ethics-the-design-and-governance-of-ethical-ai-and-autonomous-systems/

48 1000minds. (n.d.). What is human-in-the-loop (hitl) in ai-assisted decision-making?. 1000minds. Https://www.1000minds.com/articles/human-in-the-loop

49 Checkify. (n.d.). Human-out-of-the-Loop (HOOTL): Full Autonomy and the Future of AI Decision-Making. Checkify. Https://checkify.com/article/hootl/

50 Cowen, M. B. (May 2021). Is Human-On-the-Loop the Best Answer for Rapid Relevant Responses?. Joint Air Power Competence Centre. Https://www.japcc.org/essays/is-human-on-the-loop-the-best-answer-for-rapid-relevant-responses/

51 Flores, I. M. M. (2025, October 7). Chips & Brains: The Psychology of Human-AI Decision Making. Tutorials Dojo. Https://tutorialsdojo.com/chips-brains-the-psychology-of-human-ai-decision-making/

52 Wagner, B., Kuebler, J., & Zalnieriute, M. (2025, May 19). Editorial: Humans in the loop: exploring the challenges of human participation in automated decision-making systems. Frontiers in Political Science. Https://www.frontiersin.org/journals/political-science/articles/10.3389/fpos.2025.1611563/full

53 Lin, P. K. (2021, December 13). Machine See, Machine Do: How Technology Mirrors Bias in Our Criminal Justice System. Barnes & Noble. Https://www.barnesandnoble.com/w/machine-see-machine-do-patrick-k-lin/1140798895

54 Consultancy.eu. (2024, August 29). The future of decision-making: AI and human collaboration in tandem. Consultancy.eu. Https://www.consultancy.eu/news/10605/the-future-of-decision-making-ai-and-human-collaboration

55 Edwards, J. (2025, April 7). When Should Human Decision-Making Overrule AI?. Informationweek. Https://www.informationweek.com/machine-learning-ai/when-should-human-decision-making-overrule-ai-

56 Henderson, S. E. (2022, January). When Machines Can Be Judge, Jury, And Executioner: Justice In The Age Of Artificial Intelligence. Criminal Law and Criminal Justice Book Reviews. Https://clcjbooks.rutgers.edu/books/when-machines-can-be-judge-jury-and-executioner-justice-in-the-age-of-artificial-intelligence/

57 Doan, A. (2024, December 6). 9 Must-Read AI Books for Business Leaders. Nextiva. Https://www.nextiva.com/blog/ai-business-books.html

58 Plate, D., & Hutson, J.. (2025, October 27). Mind, Machine, and Will: Determinism, Responsibility, and Agency in the Age of AI. Nova Science

Publishers. Https://novapublishers.com/shop/mind-machine-and-will-determinism-responsibility-and-agency-in-the-age-of-ai/

59 Hoffman, R., & Beato, G. (n.d.). The Techo-Humanist Compass: Prioritizing Human Agency in AI Development. Porchlight Books. Https://www.porchlightbooks.com/blogs/changethis/the-techo-humanist-compass-prioritizing-human-agency-in-ai-development

60 Woods, G. (2024, September 16). The AI-Driven Leader: Harnessing AI to Make Faster, Smarter Decisions. Goodreads. Https://www.goodreads.com/book/show/216125077-the-ai-driven-leader

61 Self Publishing Titans. (n.d.). The AI-Driven Leader: Harnessing AI to Make Faster, Smarter Decisions. Self Publishing Titans. Https://www.selfpublishingtitans.com/books/detail/912

62 Abhishek. (2025, September 1). 10 Essential AI Books for Leaders in 2025. Engagedly Blog. Https://engagedly.com/blog/ai-books-for-leaders/

63 De Cremer, D. (2024, June 18). The AI-Savvy Leader: Nine Ways to Take Back Control and Make AI Work. Harvard Business Review Press. Https://store.hbr.org/product/the-ai-savvy-leader-nine-ways-to-take-back-control-and-make-ai-work/10688

64 Woods, G. (n.d.). The AI-Driven Leader: Harnessing AI to Make Faster, Smarter Decisions. AI Leadership. Https://www.aileadership.com/book

65 Evergreen, B. (n.d.). The future-solving methodology: Insights on AI and leadership. IMD. Https://www.imd.org/ibyimd/artificial-intelligence/the-future-solving-methodology-insights-on-ai-and-leadership/

66 Idoine, C. (n.d.). Gartner: The Leadership Blind Spot: Why Executive AI Literacy Will Shape Business Outcomes in 2025. The European Business Review. Https://www.europeanbusinessreview.com/gartner-the-leadership-blind-spot-why-executive-ai-literacy-will-shape-business-outcomes-in-2025/

67 Arete Coach LLC. (2025). The AI-Literate Leader: What Every CEO Needs to Know and Say About AI in the Boardroom. Arete Coach. Https://www.aretecoach.io/post/the-ai-literate-leader-what-every-ceo-needs-to-know-and-say-about-ai-in-the-boardroom

68 Chen, C. (2025, November 19). From algorithm to workforce: Preparing supply chain leaders for the AI literacy era. Supply Chain Management Review. Https://www.scmr.com/article/from-algorithm-to-workforce-preparing-supply-chain-leaders-for-the-ai-literacy-era

69 Tambe, P. B. (2025, September 30). Reskilling the Workforce for AI: Domain Expertise and Algorithmic Literacy. Management Science. Https://pubsonline.informs.org/doi/10.1287/mnsc.2022.03968

70 Singal, P. (2025, November 4). From Data to Decisions: How AI Is Redefining Executive Leadership. Intelligenthq. Https://www.intelligenthq.com/from-data-to-decisions-how-ai-is-redefining-executive-leadership/

72 Brown, S. (n.d.). When humans and AI work best together — and when each is better alone. MIT Sloan. Https://mitsloan.mit.edu/ideas-made-to-matter/when-humans-and-ai-work-best-together-and-when-each-better-alone

73 Daugherty, P. R., & Wilson, H. J. (n.d.). Human + Machine. LIT Videobooks. Https://litvideobooks.com/human-+-machine

74 Woods, G. (2024, September 16). The AI-Driven Leader: Harnessing AI to Make Faster, Smarter Decisions. Goodreads. Https://goodreads.com/book/show/216125077

75 Moreau, J. C. (2025, September 6). Leadership in the Age of AI: Human Strengths That Machines Can t Replace. Barnes & Noble. Https://www.barnesandnoble.com/w/leadership-in-the-age-of-ai-julian-cade-moreau/1148146940

76 Moreau, J. C. (2025, September 6). Leadership in the Age of AI: Human Strengths That Machines Can't Replace. Mindful Pages Publishers.

Https://www.mindfulpagespublishers.com/product-page/leadership-in-the-age-of-ai-human-strengths-that-machines-can-t-replace

77 Mellen, M. (2025, November 27). Notes from "The AI-Driven Leader" by Geoff Woods. Mickmel. Https://www.mickmel.com/notes-from-the-ai-driven-leader-by-geoff-woods/

78 Sindell, T. (2025, July 16). Leading in the AI Era: The 8 Human Skills That Will Define Tomorrow's Leaders. Skyline Group. Https://skylineg.com/resources/blog/leading-in-ai-era-8-critical-human-skills-executives-need

79 Kahan, S. (n.d.). Leadership Competencies in the Era of AI. Visionary Leadership. Https://visionaryleadership.com/ai-leadership/

80 Superhuman Team. (2025, September 17). The AI adoption roadmap for getting your entire team on board. Superhuman Blog. Https://blog.superhuman.com/ai-adoption-roadmap/

81 Fusemachines. (n.d.). AI Strategy Roadmap 2025. Fusemachines. Https://fusemachines.com/resources/ebooks/ai-strategy-roadmap-2025/

82 Levin, A. (2025, November 7). From Strategy to Execution: A Roadmap for AI Adoption in Enterprises Driving Business Transformation and Integration. Authority AI. Https://authorityai.ai/a-roadmap-for-ai-adoption-in-enterprises-driving-business-transformation-and-integration/

83 Maas, M. M. (2025). OUP book: Architectures of global AI governance. Lawai. Https://law-ai.org/architectures-of-global-ai-governance/

84 Nathaniel, R. (2021, October 24). Will the Machine Take Over? Ethics in Artificial Intelligence (Book Review). User Experience Magazine. Https://uxpamagazine.org/will-the-machine-take-over-ethics-in-artificial-intelligence-book-review/

85 Blackman, R. (2022, July 12). Ethical Machines: Your Concise Guide to Totally Unbiased, Transparent, and Respectful AI. Harvard Business Review

Store. Https://store.hbr.org/product/ethical-machines-your-concise-guide-to-totally-unbiased-transparent-and-respectful-ai/10539

86 Almog, D., Gauriot, R., Page, L., & Martin, D. (2024, December 17). AI Oversight and Human Mistakes: Evidence from Centre Court. ACM Digital Library. Https://dl.acm.org/doi/10.1145/3670865.3673481

87 mckendrick, J., & Thurai, A. (2022, September 15). AI Isn't Ready to Make Unsupervised Decisions. Harvard Business Review. Https://hbr.org/2022/09/ai-isnt-ready-to-make-unsupervised-decisions

88 Kimaru.ai. (n.d.). Decision Intelligence Recommended Reading. Kimaru.ai. Https://kimaru.ai/decision-intelligence-recommended-reading/

89 Larsen, B., & Dignum, V. (2024, October). AI value alignment: How we can align artificial intelligence with human values. World Economic Forum. Https://www.weforum.org/stories/2024/10/ai-value-alignment-how-we-can-align-artificial-intelligence-with-human-values/

91 Schuster, N., & Kilov, D. (2025, June 21). Moral disagreement and the limits of AI value alignment: a dual challenge of epistemic justification and political legitimacy. Pubmed Central. Https://pmc.ncbi.nlm.nih.gov/articles/PMC12628449/

92 Wikipedia. (n.d.). AI alignment. Wikipedia. Https://en.wikipedia.org/wiki/AI_alignment

93 O'Reilly Media. (n.d.). Principles of AI Governance and Model Risk Management. O'Reilly. Https://learning.oreilly.com/library/view/principles-of-ai/9798868809835/

94 Hoberman, S.. (n.d.). The Book Look: AI Governance. TDAN.com. Https://tdan.com/the-book-look-ai-governance/31908

95 Rosella, L. C., Shaw, J., Guha, S., Abejirinde, I. O., Gibson, J. L., Lipscombe, L., Kornas, K., Zaim, R., Chui, V., & Itanyi, I. U. (2025, July 10). A participatory approach to deploy responsible artificial intelligence for

diabetes prediction and prevention. Pubmed Central. Https://pmc.ncbi.nlm.nih.gov/articles/PMC12254626/

96 Yang, S., & Delgado, F. (n.d.). The Participatory Turn in AI Design: Theoretical Foundations and the Current State of Practice. Montreal AI Ethics Institute. Https://montrealethics.ai/the-participatory-turn-in-ai-design-theoretical-foundations-and-the-current-state-of-practice/

97 Delgado, F., Yang, S., Madaio, M., & Yang, Q. (2023, October). The Participatory Turn in AI Design: Theoretical Foundations and the Current State of Practice. ACM Digital Library. Https://dl.acm.org/doi/fullhtml/10.1145/3617694.3623261

98 Corrigan-Kavanagh, E., Plumbley, M., & Frohlich, D. (2025, August 21). Applying a Virtual World Café Method for Participatory Design of AI Systems. Journal of Participatory Research Methods. Https://jprm.scholasticahq.com/article/140957-applying-a-virtual-world-cafe-method-for-participatory-design-of-ai-systems

99 Upmann, P. (n.d.). Adaptive Governance Frameworks: Flexibility for Technological and Ethical Evolution. AIGN. Https://aign.global/ai-governance-insights/patrick-upmann/adaptive-governance-frameworks-flexibility-for-technological-and-ethical-evolution/

100 Fractal Analytics Inc.. (2024, December 13). Building resilient AI: The role of adaptive governance and cognitive compliance. Fractal. Https://fractal.ai/article/adaptive-governance-and-cognitive-compliance-for-resilient-ai

101 Reuel, A., & Undheim, T. A. (2024, June 06). Generative AI Needs Adaptive Governance. Arxiv. Https://arxiv.org/html/2406.04554v1

102 GAN Integrity. (2024, July 23). AI Governance, A Critical Framework for Organizations. GAN Integrity. Https://www.ganintegrity.com/resources/blog/ai-governance/

103 Shirkhanloo, A. (2025, February 19). Beyond compliance: The case for adaptive AI governance. IAPP. Https://iapp.org/news/a/beyond-compliance-the-case-for-adaptive-ai-governance

104 unesco. (n.d.). Ethics of artificial intelligence. Unesco. Https://www.unesco.org/en/artificial-intelligence/recommendation-ethics

105 Shah, R. (2019, October 19). Summary of Stuart Russell's new book, "Human Compatible". Effective Altruism Forum. Https://forum.effectivealtruism.org/posts/tshffdagehzoh6bzr/summary-of-stuart-russell-s-new-book-human-compatible

106 Hemphill, T. A. (2020). Human Compatible: Artificial Intelligence and the Problem of Control by Stuart Russell. Cato Journal. Https://www.cato.org/cato-journal/spring/summer-2020/human-compatible-artificial-intelligence-problem-control-stuart

107 Ingram, K. (2020). Book Review: Human Compatible: Artificial Intelligence and the Problem of Control. Ethically Aligned AI. Https://www.ethicallyalignedai.com/post/book-review-human-compatible-artificial-intelligence-and-the-problem-of-control

108 Hanson, R. (2020, August 26). Russell's Human Compatible. Overcoming Bias. Https://www.overcomingbias.com/p/russells-human-compatiblehtml

109 Alexander, S. (2020, January 30). Book Review: Human Compatible. Slate Star Codex. Https://slatestarcodex.com/2020/01/30/book-review-human-compatible/

110 mccluskey, P. (2020, January 18). Book review: Human Compatible. Lesswrong. Https://www.lesswrong.com/posts/5fr8zz4mtpvcizu4k/book-review-human-compatible

Made in the USA
Monee, IL
04 November 2026

67381308R00105